PRACTICAL STEREOLOGY

PRACTICAL STEREOLOGY

John C. Russ
North Carolina State University
Raleigh, North Carolina

PLENUM PRESS • NEW YORK AND LONDON

Library of Congress Cataloging in Publication Data

Russ, John C.
  Practical stereology.

  Bibliography: p.
  Includes index.
  1. Stereology. I. Title.
Q175.R855   1986   502.′.87   86-21253
ISBN 0-306-42460-6

© 1986 Plenum Press, New York
A Division of Plenum Publishing Corporation
233 Spring Street, New York, N.Y. 10013

Printed in the United States of America

Preface

Stereology, the study of the three-dimensional structures of specimens by examination of various two-dimensional images, is a field that is at its heart largely mathematical. However, it is being used increasingly as an important tool in such diverse fields as materials science, food science, geology, astronomy, and the life sciences.

The existing sources of information for a student interested in the use of the techniques to characterize specimens are for the most part heavily theoretical (and somewhat intimidating), yet they generally overlook subjects such as machine-based measurements. But computerized instruments are becoming common, and their operators need to know how to understand and interpret the data. Likewise, related methods such as image processing, stereoscopic measurement, and serial sectioning are often performed by the same researchers, on the same specimens, and perhaps using the same instruments. Purists may wish to exclude these from the field of stereology, yet they are at least closely related and in need of some instructional support.

The literature on stereology is extensive, yet much of it is largely inaccessible to the student. Not only is it physically dispersed in many journals covering an extensive range of disciplines, but much of it is densely mathematical, using notation that is often unusual and not explained in the article. The books in the field, which are listed in the bibliography, are either somewhat out-of-date, or organized as references rather than tutorials, or horrendously expensive. By writing this entire text (figures and all) on an Apple Macintosh and printing it on a Laserwriter, I have been able to keep the quality high and the cost moderate (and maintain close control over the entire process). The consequence is that any errors can only be my own.

The aim of this text is to provide a set of tools whose routine use can be easily learned by the working scientist and which can be taught to students at the advanced undergraduate or graduate level as they become involved in laboratory work requiring them. Rather than emphasis on derivations of complex relationships, I have tried to build a conceptual framework that ties the measurements to more practical aspects of interpretation. If the reader can learn to drive the car, and perhaps make minor repairs, that is enough; it is not necessary to know how to build a car from scratch.

The text is self-contained, assuming no specialized mathematical (or statistical) background from the student, and a minimal acquaintance with various microscopic images (either light or electron). Homework for the course is largely the measurement of various images, and can be supplemented by additional images appropriate to the discipline(s) of most direct interest to the students. If the appendix on geometric probability is included, students can be expected to create a few simple programs like those shown, but for other geometries.

I am indebted to Tom Hare for critical reviews of the material and an endless enthusiasm to debate and derive stereological relationships; to John Matzka at Plenum Press for patiently instructing me in the intricacies of typesetting; to Chris Russ for helping to program many of these measurement techniques; and especially to Helen Adams, both for her patience with my creative fever to write yet another book, and for pointing out that the title, which I had intended to contrast to "theoretical stereology," can also be understood as the antonym of "impractical stereology."

John C. Russ
Raleigh, NC
July, 1986

Contents

Chapter 1

Statistics

Stereology is defined (we will refine and expand upon this definition in subsequent chapters) as the study of the three-dimensional structure as revealed in two-dimensional images, usually of sections through it. Stereometry is, therefore, the measurement of these structures. As in most measurements which we can perform in the sciences, there are errors arising from bias in our techniques or finite precision in our methods which require statistical procedures to analyze. However, in stereology we are forced by the very nature of the measurement process to sample the true 3-D structure, and the extent to which our samples reveal the "truth" about what is there is fundamentally a statistical problem. Consequently, and because it is my experience that many students are not familiar or comfortable with the basics of statistics, it is necessary to begin with an introduction. This is not intended as a complete course in statistics (!), but it should provide the fundamentals needed for the specific application to stereometric measurements. Armed with these statistical concepts and tools, we will be able to evaluate and compare the measurements to be introduced in the following chapters.

Accuracy and Precision

In some types of measurements, for instance applying a ruler to a brick to determine its length, repeated measurements give slightly different results (whether performed by a single individual or by several different ones, the latter usually showing even more variation). This kind of "error" is not due to moral turpitude or original sin, but to the fact that measurements are of finite precision. The variation, which we will soon characterize as a "standard deviation," has a real number value (that is, with a decimal fraction) in the same units as the measurement which it describes. For instance, we might describe the length as 6.15 ± 0.072 inches.

A somewhat different situation arises when we are counting things. For instance, if you were asked to count the number of cars passing under a highway overpass during a 10 minute period, it would seem that no error should be present (assuming you don't fall asleep, or mistake a truck for a car, or some overt error which we would classify as bias in the measurement). But now consider that the value you obtain, which is an integer (rather than a real number), may be asked to represent the expected number of cars which would pass under the bridge in another similar time period. You will realize that the number you counted will not exactly match that for another period, and perhaps you may want to count more periods to get a better average (we will call this the mean). But still you can only achieve a best estimate of the expected rate for a typical time. This kind of error is called sampling error, and it is the type we shall normally encounter in image measurement, because most of the things that we do involve counting, and clearly the images we see (which may represent a combined area of a few square millimeters for light microscope images, or a few square micrometers for electron microscope images) represent only a small sample of the real structure.

To get the flavor of this sampling error, and specifically to see it in relation to a real stereometric operation, consider the metal microstructure shown in Figure 1. A standard method (which we will develop in a later chapter) for determining the grain size of this microstructure involves drawing lines and counting intercepts (that is, the number of times that the lines cross grain boundaries). Without further introduction, try the following experiment: Draw "random" lines on the image (or better, on a copy of it) amounting to a total of 1 centimeter of actual length (take into account the indicated magnification of the image). This will require drawing a number of lines, at various angles and locations. Now count the number of intersections. The "true" value – that is, the average based on many measurements including those on other, equally representative sections through the microstructure – is 125.4 intersections per centimeter of line.

How well does your answer agree with the "truth," or even with other answers from your classmates? How can you estimate whether your answer is significantly different (in a statistical sense) from the given answer? These are questions which we must use statistical means to answer.

It is likewise impossible to overlook the contribution of bias to the accuracy of results. The late George Moore, of the U. S. National Bureau of Standards, devised the following simple test (Moore 70), to which he subjected many professional metallographers and would-be stereometricists. Please do as instructed.

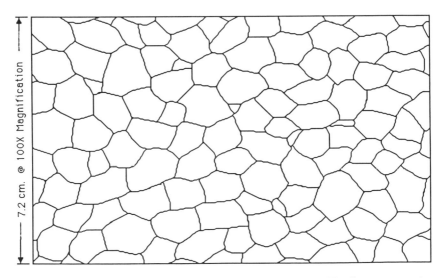

Figure 1. A "standard" microstructure for ASTM grain size 4. The lines represent grain boundaries on a flat polished section through the structure.

Recognition and Counting

Nearly all laboratories where there is occasion to use extensive manual measurement of micrographs, or counting of cells or particles therein, have reason to note that certain workers consistently obtain values which are either higher or lower than the average. While it is fairly easy to count all the objects in a defined field, either visually or by machine, it is difficult to count only objects of a single class when mixed with objects of other classes. Before impeaching the eyesight or arithmetic of your colleagues, see if you can correctly count objects which you have been trained to recognize for years, specifically the letter 'E'. After you have finished reading these instructions, go back to the beginning of this paragraph and count all the appearances of the letter 'e' in the body of this paragraph. You may use a pencil to guide your eyes along the lines, but do not strike out or mark E's and do not tally individual lines. Go through the text once only; do not repeat or retrace. You should not require more than 2 minutes. Please stop counting here!

Please write down your count on a separate piece of paper before you forget it or are tempted to "improve" it. You will later be given an opportunity to compare your count with the results obtained by others. Thank you for your cooperation!

There are several obvious errors that can be expected if the directions are not carefully followed, such as including the title or the second paragraph. There is some potential uncertainty about capital E's – should they be counted, or not, or should ONLY capital E's be counted? What should one do about a misspelling that omits an 'e'? This is a little like encountering a gap in a grain boundary in the first example. If you "know" the boundary is there, but it is not visible at the expected point of intersection (for instance due to an etching or polishing defect), should it be counted anyway?

In addition to these problems, there is the simple counting error associated with imperfect recognition. One person taking the test reported that he had a hard time remembering to count the silent e's! The histogram of results shown in Figure 2 demonstrates a substantial measurement error even in this case where there seems to be no excuse for it. And beyond this error, even if the result was accurate for this paragraph, would we be able to claim that it was representative of all of George's writing, or all technical writing, or all of English in general? These problems of generalization are at the heart of stereology.

If the images we examine are not truly representative of the structure, it is like counting the cars (in the earlier example) at 6AM (because it is convenient to do so) and hoping the results describe the traffic at noon. This kind of sampling bias cannot be dealt with or corrected using stereometric techniques. We will be assuming that the images are truly representative (in a statistical sense) so that the average we obtain does properly describe the expected results anyone else would obtain on the same material, but with different samples. Underlying this restriction is the notion of a "random" sample, that is, one that does not have any predictable relationship to the structure of the material. If the "true" microstructure has preferred orientation, we must look at many different orientations; if the surfaces are different from the interior, we must sample representative amounts of each; if there are both very large and very small features present, we shall have to use magnifications that reveal both.

A word of caution here: we have not tried to rigorously define "random" but instead will rely on the reader's intuitive notion of this important concept. It is true that random sampling will ultimately cover an entire specimen, as would uniform sampling. However, the latter would do so before any area was examined twice, whereas with random sampling, long before we saw all areas we would have begun to repeat ourselves. Random sampling is an efficient and effective means of collecting data only when it sparsely samples the entire population, or in other words when we look at a very small fraction of the possible cases.

We will observe the same thing in using random numbers to perform "Monte-Carlo" samples of geometric probability. A uniform distribution of numbers would be equivalent to performing a continuous integration or summation, and must be complete to accurately sum up the system response, whereas the use of random numbers should always predict the mean answer, but with precision that improves as the number of samples increases. This will become more evident when this sampling technique is described in the next chapter.

Random sampling is not the only possible approach. The alternative is uniform or ordered sampling, in which decisions must be made about placement of sections and fields to be measured or counted. For some types of heterogeneous specimens (especially biological ones) this is necessary to cover the various components within the specimen, but it carries with it the potential for serious bias in the results, if the sampling is not proper.

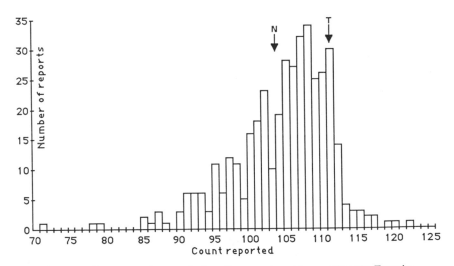

Figure 2. Representative results from the experiment in counting e's. T marks the true answer, and N the mean (average) of many trials by different persons.

The mean and standard deviation

To determine how much variation to expect in a measurement, we can make a series of independent measurements and see how much they vary. The average value of all the individual measurements is the *mean,* and is our best estimate of the value we would obtain if we repeated the measurement again. In our standard notation, the mean will be designated by μ and is calculated from the N individual x_i values as

$$\mu = (\Sigma x_i) / N$$

Barring bias in our measurement, the mean is also the best estimate of the "true" value of the measurement. These two statements are not the same! In fact this is the difference between precision and accuracy. Precision describes the reproducibility of a measurement and accuracy the degree of agreement with some objective truth determined outside our experiment. For instance, precision is the ability of your car's speedometer to read very close to the same value, say 55 mph, whenever your car is going at a particular speed, while accuracy (or inaccuracy) is the difference between the reported 55 mph reading and the truth, which might be 62 mph reported by the state trooper's radar (which for our present purposes we will take to be error-free). Accuracy can only be assessed if we can either obtain independent certification of the truth, measure known standards, or identify all of the sources of bias in our measurement. This is no easy task. We will usually discuss our results in terms of precision. By convention, we will report results with the number of significant digits shown to indicate the expected precision.

The *variance* of a series of measurements is defined as the mean of the squares of the differences between the mean value (as described above) and each of the individual values. By using the squares of the differences, the sign becomes unimportant, and so this quantity reflects how badly scattered the individual measurements are. Notice that if we make only a single measurement, it is automatically the best estimate of the mean, but we cannot estimate the amount of scatter.

In the limit, where we measure many, many individual data points, the mean value of our observations approaches the true mean of the population which is being sampled. Similarly, from the variance we can determine the amount of spread that is really present in the data, and hence the precision (and perhaps the accuracy) of the measurement. For this we use the *standard deviation,* which is defined as the square root of the variance. It would be written as

$$\sigma = \{ \Sigma [x_i - \mu]^2 / N \}^{1/2}$$

Note that sigma has the same units as the measurement value itself. Sometimes it is useful to express the precision relative to the actual value, and this *relative standard deviation* is just σ / μ (if multiplied by 100 and expressed as percent, this may be called the coefficient of variation). When we have a small number of measurements from which to estimate the mean and standard deviation, we must take into account the finite size of our sample. This is done by calculating the standard deviation of our sample (as opposed to that for the population, which would be reached in the limit) as

$$\sigma = \{ \Sigma [x_i - \mu]^2 / (N - 1) \}^{1/2}$$

The $(N - 1)$ takes into account the loss of one independent piece of information when we calculate the value of the mean. This is described in statistical parlance as "using up one degree of freedom." The number of *degrees of freedom* is the number of data points we collect less the number of derived parameters we obtain by calculation from them. In a perhaps more familiar example, solving simultaneous equations, we must have at least as many data points as we have equations. If there are more, the overdetermined set of equations can be solved for the best answers. Consider as an example the linear equation $y = Mx + B$ for a straight line. If this is fit by the familiar least-squares method to a series of x,y points to determine the "best" values of M and B, a minimum of two points are needed, and the more we have beyond that (the more degrees of freedom), the better the fit.

The standard deviation as described above is a measure of how likely you are to observe a particular value in an individual measurement, depending on how different it is from the mean value. It is also possible to estimate the standard error of the mean itself as $\sigma / \sqrt{N}$.

In the example cited earlier for measuring the length of a brick, there is no relationship between the variance or standard deviation in our measurement and the magnitude of the mean value. This is our usual experience with measurements of this type, where the precision of our measurement process depends on the tools, our skill, and the degree of control we have over external influences. But in the case of counting, it is possible to directly estimate the variance from the number of things that we count. This is a fundamental relationship whenever "random" events are counted to determine the expected frequency of occurrence. The standard deviation in a number of counts N is just $\sqrt{N}$.

For example, if we return to the example of counting cars, the precision in our estimate of the mean frequency of cars (perhaps expressed as cars per minute) depends on the number of cars we count, regardless of how long it takes us to do it. So if we count 100 cars in 2 minutes, the standard deviation is the square root of 100, or 10, and the frequency would be expressed as 50 ± 5 cars per minute. But if we were to count for 20 minutes, the count would be about 1000 cars ("about" because there is a random variation expected in the result; this example assumes that we are counting a truly random event like the emission of X-rays from a radioactive source; for the moment try to imagine that cars on the freeway are really random, too). The standard deviation would be the square root of 1000, or 32, so the mean frequency would be 1000/20, or 50 cars per second plus or minus 32/20 or ± 1.6 cars per second. Notice that the longer we count, or the more counts we obtain, the better our precision (the more accurately we can estimate the "true" mean). However, because the precision improves only as the square root of the number, increasing amounts of effort are needed to gain better results. An improvement of 2 times in precision requires making 4 times as many observations, and so forth. Figure 3 shows this relationship.

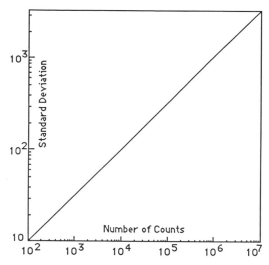

Figure 3. Relationship between standard deviation and number of counts.

In making stereometric measurements, we will encounter both types of precision: errors in measurement that are real numbers and independent of the magnitude of the number, such as the length or area of features; and errors in counting statistics, such as the number of features or the number of intersections of a line with grain boundaries, where the precision is determined directly by the number of counts.. It will be important to recognize these two different types of errors, because we can improve the latter one by obtaining more data, but this may not help the former.

Distributions

The key to understanding how measurement precision (or error) affects the expected result from a measurement is to make a number of independent measurements of the same thing, such as the number of intercepts per centimeter used to measure the grain size of a metal, and construct a plot of the frequency with which various values are obtained. This is called a probability distribution curve. Note that the measurements must indeed be of independent samples (in this case, different areas of the sample surface, or even different sections through the microstructure). The parent population for such a distribution curve might look like Figure 4. The mean μ is marked on the curve. For the curve shown, which is not symmetrical, there are two other values also marked, the most probable value (the highest point on the curve) and the median (the point that divides the distribution in half, into two parts with equal area). The median is rarely useful as a statistical parameter, but the most probable value can be important, and it is also noteworthy that it may not be the same as the mean determined by many repetitive measurements.

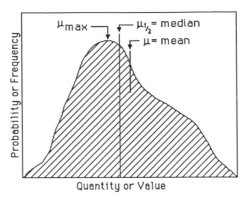

Figure 4. A representative distribution curve, showing the mean, median and most probable values.

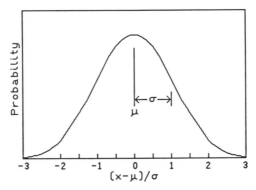

Figure 5. The normal or Gaussian curve, showing the mean and standard deviation.

What the curve shows is the probability that any particular value will be obtained when the measurement is performed. The vertical axis may be labelled probability, or frequency, or simply number of observations. For many real measurement cases, the shape of the curve is symmetrical, and furthermore has a particular shape, called the *normal* curve (or the Bell curve, or the Gaussian curve). Figure 5 shows this classic distribution.

Note that the mean is the most probable value for this distribution. Several probability distributions may be Gaussian in shape, with the same mean, but with different breadths (see Figure 6). The standard deviation (marked in the figure for each curve) is a unique descriptor of the width of the curve. Mathematically, the shape of the Gaussian or normal curve is given by

$$P_{Gaussian}(x,\mu,\sigma) = 1/[\sigma(2\pi)^{1/2}] \cdot \exp\{-((x-\mu)/\sigma)^2/2\}$$

This is a continuous function with unit total area under the curve, so the probability of observing any particular value of x (the measurement value) is calculated as a function of its deviation from the mean $(x - \mu)$ and the distribution's standard deviation σ. Furthermore, the nature of this distribution is such that the area between the points at plus and minus the standard deviation is always 0.68 (68%), between the $\pm$ 2 sigma points is 0.95 (95%), and so forth. This is why we say that it is 95 percent probable that any particular measurement will yield a value within two standard deviations of the mean, that it is 99% probable that it will yield a value within three standard deviations of the mean, and so on. Sometimes the normal curve's width is described by the "full width at half maximum" (FWHM) instead of the standard deviation. Again, the shape of the curve is fully determined and it is possible to directly relate the FWHM to sigma (it is just 2.3548 times sigma). There are many tables and computer routines that will estimate probabilities using the normal curve.

Since the shape of the curve is fully determined by the mean and standard deviation, changing these parameters can produce an entire family of curves. Figure 6 shows several with different means, standard deviations, and heights (or areas). In practice, we rarely see the continuous form of this curve. It is usually convenient to collect data into bins based on finite ranges of value, and to plot the number of observations in each bin. This gives rise to the histogram form of the curve, as shown in Figure 7.

When counting statistics apply, we saw before that the mean and standard deviation are related (sigma is the square root of the mean). So in this case, the shape of the normal curve is fully determined by a single value, the mean μ. If the mean value is, for instance, 100 counts, then the standard deviation is 10 and 95% of the time while making repeated independent counts we would expect to obtain a value between 80 and 120 (plus or minus two standard deviations). Of course, in this case it is not really proper to speak of a continuous smooth probability curve. There is no way we can obtain a fractional count value, so the curve is really a histogram with discrete values, where the height of each column in the plot times its width gives the same area (and hence the same probability of occurrence) as the area under the

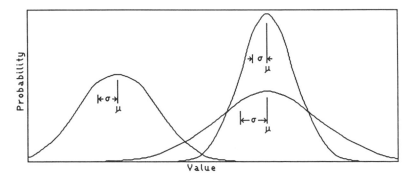

Figure 6. Several normal (Gaussian) curves with different means and standard deviations, and different relative probabilities (heights).

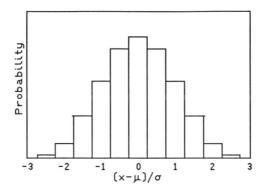

Figure 7. The normal (Gaussian) curve shown in histogram form.

corresponding portion of the continuous normal curve. This is a correct description of the probability of obtaining any particular integral number of counts.

The normal curve is a good model for the expected statistical probability of various errors when counting relatively large numbers of events, but it does not apply to small numbers. This is easy to see by considering that the curve is symmetrical and has very low but broad tails that reach out infinitely far on both sides. For example, for a mean of 4 counts, the standard deviation is 2 (the square root of 4). This would imply that there was a 95% probability of obtaining a count between 0 and 8, with a 2.5% probability of getting a value greater than 8 and a 2.5% probability of getting a value less than zero. This is, of course, meaningless in a counting situation. Actually, when counting things, the exact description of the probability curve is Poisson rather than Gaussian. For modestly large numbers the two agree exactly, but for small numbers the Poisson curve is asymmetric and does not permit negative values. Figure 8 shows an example.

For the Poisson distribution, the standard deviation is always determined as the square root of the mean, so it takes only one parameter to define the curve. The

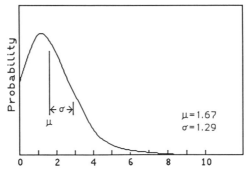

Figure 8. A Poisson distribution, with its mean and standard deviation.

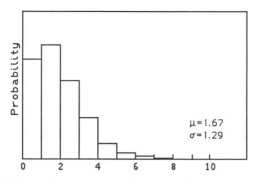

Figure 9. A histogram of the Poisson distribution in Figure 8.

mathematical expression for the shape of the curve (or histogram, as we are dealing with a discrete distribution with integral x values) is just

$$P_{Poisson}(x,\mu) = (\mu x / x!) \cdot e^{-\mu}$$

where the ! symbol indicates factorial, and μ is the mean. Note that for this distribution, there may be a finite probability of counting zero events, and the mean is larger than the most probable value (and may not be an integer). When the mean is large, the Poisson distribution is indistinguishable from the Gaussian distribution with the standard deviation fixed as the square root of the mean. The Poisson distribution is normally shown as a histogram rather than a continuous curve, to emphasize the fact that only integer values are possible in counting (Figure 9).

Many real physical measurements have distributions that correspond, either exactly or closely enough to be a useful measurement tool, to the Poisson or Gaussian shape. A variant that is also frequently observed is the *log-normal* distribution, which often turns up in particle sizes. This is just a plot (or histogram) of measurement probability in which the horizontal axis is the logarithm (to any base) of the measurement value (a linear dimension like length, or a volume or surface area, for instance), and for which the shape is like the Gaussian or normal curve. Figure 10 shows a log-normal distribution, and Figure 11 shows how it would appear if plotted on a linear scale.

There are alternative ways to plot such data that make them easy to interpret. For instance, using probability paper, a log-normal distribution plots as a straight line as shown in Figure 12 (in fact, probability paper is designed specifically to make this happen). This kind of plot is sometimes used in sieving and other measurement techniques, simply because the practical incidence of log-normal distributions is common.

For these types of curves, the measured parameter is a real number (not a count), and so the mean and the standard deviation of the curve are not related. Furthermore, the standard deviation has units of the log of the measurement dimension, and so represents a ratio of size rather than a linear size range. This concept will be expanded later on.

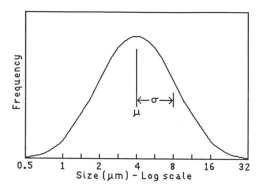

Figure 10. A Log-normal size distribution curve.

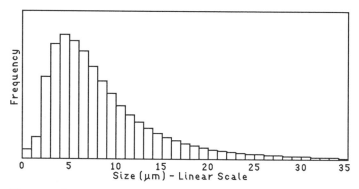

Figure 11. The data from Figure 10 plotted as a histogram on a linear scale.

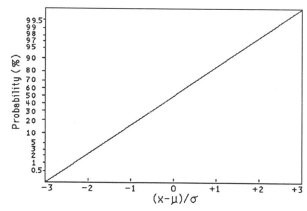

Figure 12. The log-normal distribution plotted on probability paper. The vertical axis is scaled in terms of standard deviation, but marked in units of probability based on the Gaussian (normal) curve.

There are other transformations besides the logarithm of a value that are also useful. For instance, if area is measured, a plot of number versus the square root of area may show a more symmetric shape than a plot of number versus area. In this example, the square root of the area is directly proportional to a linear measure of size, for uniformly shaped features.

Comparisons

When a series of measurements have been made, and the data accumulated into a distribution curve or histogram, we shall want to compare the results to a standard shape (like the normal or log-normal curve) to see whether this simple description can be applied to our data. Also, comparisons between distributions measured on different samples are important. There are several statistical tools available for these tests.

One way to compare two distributions, or a measured distribution to a theoretical one such as a normal curve, is the chi-squared test. Chi-squared is statistical jargon or shorthand for the process of summing up the squares of the differences between the two sets of data. This sum is what is minimized in so-called "least-squares" fitting techniques. As an example, let us see how to compare a measured distribution to a normal (Gaussian) distribution with the same mean and standard deviation as the measured data. In this situation, we will sum the value over all bins, for the differences between the observed frequency (number of observations) in each bin with the value predicted by the normal distribution.

$$\chi^2 = \Sigma (P_i - G_i)^2 / P_i$$

where for each bin i, P is the measured frequency (or number of observations), and G is the Gaussian probability predicted by

$$G_i = N \, \exp \left(- \left((x_i - \mu)/\sigma\right)^2 / 2\right)$$

where N is the total number of observations, x_i is the value of the center of the measurement bin, and μ and σ are the mean and standard deviation of the distribution, as usual. (Note: using the Gaussian probability for the value at the center of the measurement bin, instead of integrating across the bin width, introduces a small error which is ignored here.)

The value of chi-squared characterizes the differences between the observed and expected frequencies. The numerator of each term measures the difference, and the denominator expresses the expected difference (remember that the standard deviation in a number of counts, in this case the number of observations in a particular bin based on measurement value, is expected to be just the square root of the number, so the denominator is in effect the square of the expected standard deviation in the height of that bin in the histogram). If the observed distribution exactly follows the theoretical one, the value of chi-squared will be zero. In order to compare the values to tables for chi-squared, it is common to use the "reduced" chi-

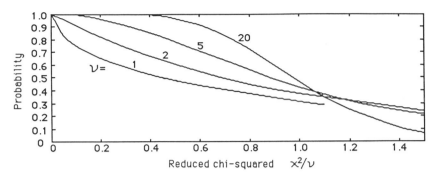

Figure 13. The probability of exceeding an observed value of
reduced chi-squared for various degrees of freedom ν.

squared value obtained by dividing by the number of degrees of freedom ν, which in
this case is just $n-2$ where n is the number of bins in the distribution.

Figure 13 shows the probability of exceeding an observed value of chi-squared for various values of the reduced chi-squared value and number of degrees of freedom. More complete tables can be found in most statistics texts. Notice that the result is expressed as a probability.

In the example histogram in Table 1, the actual number of observations are listed with the value predicted by a Gaussian distribution having the same mean (4.2) and standard deviation (1.59) as the observed data. The total chi-squared value is 5.01 for seven bins, with $5 = 7 - 2$ degrees of freedom. The figure shows that the probability of exceeding this reduced chi-squared value (1.0) by chance is 45%, so we would conclude that a normal distribution is a reasonable fit to the histogram.

A much simpler and faster comparison between different distributions is afforded by the Analysis of Variance, or *ANOVA* test. This just compares the means of the distributions, and sees whether the difference between them is larger than would be expected based on their respective standard deviations. This technique is particularly easy to apply, and also allows more than two distributions, or ones with

Table 1. Example for chi-squared test of normality

Bin No.	Observations	Predicted	$(P-G)^2/G$
1	2	2.359	0.055
2	10	6.846	1.453
3	11	13.392	0.427
4	17	16.906	0.012
5	14	15.558	0.187
6	8	10.853	0.216
7	7	3.813	2.664
		Total	5.014

different numbers of observations and distributions with different measurement bins, to be compared to see whether they could all represent the same parent distribution.

To perform the *ANOVA* test, we must calculate the amount of the variation within the measurements within each set, as compared to that between sets. This is expressed as a factor

$$f = \{SSA \mid n_1\} \mid \{SSE \mid n_2\}$$

where *SSA* is the sum of squares of differences within a class (or, as it is often called in this field, a "treatment" - jargon reflecting the fact that analysis of variance is often used in evaluating medical and sociological data), and *SSE* is the difference between the total sum of squares of differences (*SST*) and the *SSA* value. These are individually written as

$$SST = \Sigma\Sigma (y_{ij} - y_{mean})^2$$

$$SSA = \Sigma\Sigma (y_i^* - y_{mean})^2 = \Sigma n_k \cdot (y_i^* - y_{mean})^2$$

$$SSE = \Sigma\Sigma (y_{ij} - y_i^*)^2 = SST - SSA$$

The summation for the *SST* value proceeds over each class of observations (there are k classes each with n_k data points), and each value within each class, while the summation for *SSA* covers only the means of the k classes. y_{ij} is the value of observation j in class i, y_i^* is the mean value for class i, and y_{mean} is the global mean (average of all the values) which can also be obtained by summing the individual class means times the number of observations in each class, and dividing by the total number of observations.

The f value, along with the two degrees of freedom $\{v_1 = k - 1; \; v_2 = t - k\}$ where t is the total number of observations in all classes, can then be compared to values from Figure 14 (or extensive published tables in most statistics texts, which include other confidence limits) to see whether the value exceeds that for the stated probability of significance.

Table 2. Example data for ANOVA test

Set No.	Observations	Mean
1	11,17,14	14.00
2	15,17,10,13	13.75
3	9,12,14,8,17,15,11	12.285
4	12,10,16,7,16	12.20
5	15,13,11,9,10,14,11	11.857
6	11,15,12,12	12.50

$$SST = 223.367$$
$$SSA = 16.531 \quad \text{Global Mean} = 12.567$$
$$SSE = 206.836$$

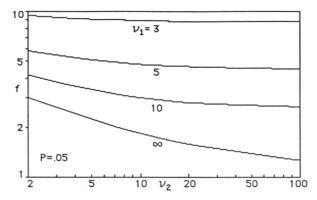

Figure 14. Critical values of *f* for various degrees of freedom,
for a probability of significance of 95%.

For instance, for the 6 sets of data with a total of 30 observations shown in Table 2, the number of degrees of freedom would be 5 and 24. The graph shows that a value of *f* greater than 4.5 would be expected only 5% of the time from chance alone. For these data, the value of *f* is 0.38, less than 4.5, so we would conclude that at this level of confidence, there is no indication that the data in the individual data sets come from different parent populations

The chief difficulty with the analysis of variance test is that it makes the implicit assumption that the data have a normal distribution within each class. While this is often the case with physical measurements, it is not always so. It is always possible to calculate a mean and standard deviation, but if the data are not normally distributed, these two parameters do not fully describe the distribution, and hence cannot be used to compare different ones.

Some insight into the shape of a distribution can be obtained by calculating the skew and kurtosis, along with the variance (or standard deviation). Just as the variance is a second moment of the distribution (the sum of the squares of the deviations), the skew is the third moment and the kurtosis is the fourth moment. For a set of *n* observations, they would be calculated as

$$Skew = m_3 \, / \, (m_2)^{3/2}$$

$$Kurtosis = m_4 \, / \, (m_2)^2$$

where

$$m_k = \Sigma \, (x_i - x_{mean})^k \, / \, n$$

A normal (Gaussian) distribution will have a skew of zero and a kurtosis of 3. The sign of the skew indicates the direction in which the distribution is asymmetric (a negative sign means the data trail off more gradually below the mean, and conversely). Values of kurtosis larger than 3 result from distributions that are more sharply peaked than the normal curve, while lower values correspond to distributions that are more flat-topped. Some comparison of distribution shapes can be carried out with these parameters in specific instances.

A more universal approach is to use "nonparametric" tests to compare data sets, and the most common of these is the Wilcoxon test (in the version presented here, this is sometimes called the Mann-Whitney or *"U"* test). The name "nonparametric" means that the test does not assume any particular shape for the distribution of the observations. To apply the procedure to two sets of data, the individual values from both sets are first ranked in value order, and each one is assigned a number from *1* to *T* (the total number of observations, n_1+n_2, in both sets). In cases where several observations have the same value, the rank of all is set to the average rank (e.g. if the 7th through the 10th observations are equal, each would have a rank of 8.5). Then the sum of ranks of observations in each set is calculated and denoted by W_1 and W_2 (by convention, W_1 is the smaller sum of ranks; since the sum W_2+W_1 depends only on the total number of observations, only W_1 is really needed). The test value *U* is calculated from W_1 and the number of observations in set 1, and reflects the number of ways that the observations could be arranged to give the same value, by chance.

$$U = W_1 - n_1(n_1 + 1)/2$$

This is compared to the critical values shown in Figure 15. If the value exceeds that shown for probabilities of 1% and 5%, respectively, the difference between the two data sets is not significant at that level (i.e. it could have occurred by chance).

Table 3 shows an example, with two sets of observations containing 10 and 12 values, respectively. By listing them in order, the ranks of the observations in set 1 are evident. The actual values of the observations are unimportant. The calculated value of *U* is 57, which is greater than the value of 30 on the graph. Hence, the data are not considered to be significantly different (they could have come from the same population).

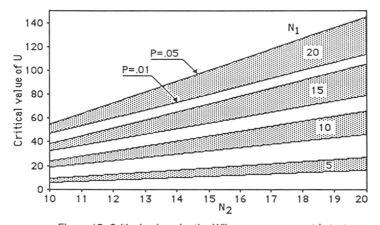

Figure 15. Critical values for the Wilcoxon nonparametric test.

Table 3. Example data for Wilcoxon test

Set 1	27	29	30 31 31	32				36 37	41 43	($n_1 = 10$)
Set 2		28	30 31 31 31	33	34	34	35	37 39 40		

Rank (set 1)	1	3	4.5 8 8	11	16 17.5	21 22 ($W_1 = 112$)

$$U = 112 - (10 \cdot 11 / 2) = 57$$

Correlation

These comparisons between sets of observations have ignored the order of data within each set. In some cases, we may wish to compare individual values (for instance, different parameters such as shape and size for individual features). This is most commonly done using a correlation coefficient. If we plot each observation as a point on a graph, with each variable assigned to one axis, we obtain a plot from which the tendency of the two values to vary together can be estimated. The correlation coefficient measures this tendency, and is in effect a measure of how closely the data cluster about a straight line (correlations using other functional relationships are also possible, but we will restrict the discussion here to the linear case).

For a series of N data pairs X_i, Y_i, an equation of the form $y = mx + b$ can be fit by determining the values of m and b which give a best fit (in the least-squares sense of minimizing the deviation of the points from the line) by calculating

$$m = \{ N \, \Sigma x_i \, y_i - \Sigma x_i \, \Sigma y_i \} / \{ N \, \Sigma x_i^2 - (\Sigma x_i)^2 \}$$

$$b = \{ \Sigma y_i - m \, \Sigma x_i \} / N$$

This gives the dependency of y upon x; in other words, the deviations which are minimized in the least-squares sense are the vertical (y) differences between the points and the line. In general, performing the fit the other way (minimizing the differences in the x-direction) will produce a different result. Only in the case of perfect correlation (the points lie exactly on the line) will the two fits be the same. If we use m' as the slope for the fit $x = m'y + b'$, then the correlation coefficient r is defined as $\sqrt{(m \, m')}$. It can be calculated as

$$r = \{N\Sigma x_i \, y_i - \Sigma x_i \, \Sigma y_i\} / \{(N\Sigma x_i^2 - (\Sigma x_i)^2\}^{1/2} \{N\Sigma y_i^2 - (\Sigma y_i)^2\}^{1/2}$$

The magnitude of r can vary from 0 (no correlation at all) to 1 (perfect correlation). Figure 16 illustrates cases with high and low degrees of correlation. The sign of the coefficient may be either positive or negative; negative values simply reflect the fact that an increase in x is accompanied by a decrease in y, and vice versa.

The significance of a value for r depends upon the number of degrees of freedom, which is $n - 2$ where n is the number of data points. This is shown in

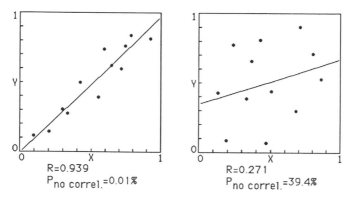

R=0.939
$P_{no\ correl.}=0.01\%$

R=0.271
$P_{no\ correl.}=39.4\%$

Figure 16. Data sets with high and low correlation coefficients.

Figure 17; the values of r for a particular number of degrees of freedom can be considered probably significant (at the levels shown) if they exceed the values marked for the various curves. For instance, a value of 0.6 for 20 observations would be significant at the 0.01 level (it could arise by chance only 1% of the time).

There are variations of the correlation test in which the rank orders of the two variables are used, rather than their actual values. Also, sometimes we plot and try to correlate using transformed values, such as the log of size or the cube of a linear dimension. When this is done, the fundamental assumption of the regression (that all deviations of points from the line should be equally minimized) may be wrong. In either case, the general meaning and interpretation of the correlation coefficient remains the same but the significance of particular values will be altered.

Nonlinear fitting

In many practical cases we shall encounter, the relationship between two variables is not linear . See for instance the example of Figure 16 in Chapter 6. The linear fit of form factor against size produces a correlation coefficient with a

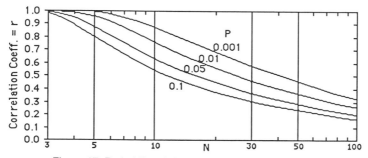

Figure 17. Probability of significance of values of the linear correlation coefficient r for N observations.

magnitude of 0.76; the sign is negative, reflecting the fact that form factor decreases with increasing size. For the number of particles in the plot (264) this means that the result is very highly significant. However, the correlation coefficient tells us that the linear relationship we have computed, of the form $y = mx + b$, accounts for only three-fourths of the variation in the data. Part of the rest is due to the scatter of the points about the line. But visual examination of the plot suggests that the trend of the data points is not really quite linear. The data appear to follow a curve that is flatter at small sizes and steeper for large ones.

When these kinds of nonlinear relationships appear, or are suspected from other knowledge about the materials, linear regression will give only partially useful results. However, there are related techniques that allow nonlinear models to be fit to the data. The simplest of these is to use polynomial terms to allow the fit to follow curves in the data. The general form of the equation is

$$y = b_0 + b_1 x + b_2 x^2 + \dots$$

and as in the case of linear fitting, we proceed by a desire to find the values of the b coefficients that give the best average agreement between the measured and calculated values of y, which means that we want to minimize chi-squared

$$\chi^2 = \Sigma (\delta y_i / \sigma_i)^2 = \Sigma [(y_i - f(x_i))^2 / \sigma_i^2]$$

where the $f(x)$ function is the polynomial (or other function, as we will see shortly), and σ is the standard deviation of each individual data point, over which the summation is carried out. Usually it is acceptable to assume that all of the data points have equal precision, and the σ terms cancel out in the minimization. The process requires that we take the partial derivative of chi-squared with respect to each coefficient, which gives a set of simultaneous equations that are solved by matrix methods. The procedure is somewhat beyond the scope of this text, except to note that the resulting inverse matrix contains information that describes the quality of the fit.

A similar procedure is followed when multiple linear regression is carried out. In this case, the equation is

$$y_i = b_o + b_1 x_i + b_2 z_i + \dots$$

where the dependent variable y is described as a function of several independent variables x, z, etc. An example would be to look for a functional dependence of size on aspect ratio (or some other shape factor) and angular orientation. The matrix procedure for solving the equations is identical to that for polynomial regression.

For the case of multiple regression, a correlation coefficient is defined that behaves much like the linear correlation coefficient we have already encountered.

$$R^2 = \Sigma b_j (\sigma_j / \sigma_y) r_{jy}$$

where the summation is over all of the independent variables, the b terms are the coefficients for each variable, the r terms are the ordinary coefficients of linear correlation as defined earlier, and the σ values are the standard deviations of the data for each variable, calculated in the usual way as

$$\sigma_j^2 = \Sigma (x_{ji} - x_{jmean})^2 / (N-1)$$

The r_{jy} terms are available directly in the inverse matrix. They can be examined individually to determine the relative contribution of the various independent variables to the overall fit of the model selected. The R^2 value can be examined both to determine the significance of the fit (using the same test with reduced chi-squared and the number of degrees of freedom as described earlier for the linear case), and also as a measure of the portion of the variation in the y (dependent variable) values that is accounted for by the model.

There are other nonlinear regression models that are commonly encountered in work with feature size data. For instance, consider the log-normal distribution already cited. When the logarithm of the size is substituted for size in the linear regression model, we are really making it into a nonlinear problem. The difficulty arises because usually in obtaining the best-fit coefficients, the chi-squared calculation assigns equal weights to each of the data points. But when they have been transformed, for instance by a log function, this is no longer valid. This is the case whenever nonlinear functions are used, including the derivation of formfactors, volume and surface area, and most of the other derived parameters that have been introduced.

Consider as a simple example the model

$$y = a \cdot bx$$

where a and b are constants. This can be simplified to a "linear" expression of the form

$$\log y = \log a + x \log b$$

but the chi-squared summation becomes

$$\chi^2 = \Sigma [(\log y_i - \log a - x_i \log b)^2 / \sigma_i^2]$$

In effect, the nonlinear substitution of values has also changed the relative uncertainties of the data points, as shown in Figure 18.

The two plots of the function, in linear and logarithmic coordinates, show the representative error bars assuming the original y values had equal errors. Note the vast difference in the error bars for $\log(y)$. To compensate for this trend, we should modify the expression used for σ in the usual expression for chi-squared.

$$\sigma_i' = d (\log y_i) / d y_i = \sigma_i / y_i$$

This, and other weighting functions for other transform models, can significantly alter the coefficients obtained by fitting of model equations when derived variables are used in the fits.

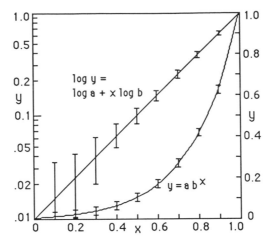

Figure 18. Plots of $y = ab^x$ on linear and log scales, showing the change in the error bars (assumed equal in the linear plot).

The modification of the weighting factors is closely related to the problem of propagation of errors. When some functional relationship has been derived or proposed for a dependent variable y in terms of independent variables x and z (for instance), the variance (square of the standard deviation) in y depends on the variance for the independent variables as

$$y = f(x,z)$$

$$\sigma_y^2 = \sigma_x^2 (\partial y/\partial x)^2 + \sigma_z^2 (\partial y/\partial z)^2 + 2\sigma_{xz}^2(\partial y/\partial x \cdot \partial y/\partial z)$$

where the last term drops out when x and z are uncorrelated variables. These calculations can be made straightforwardly, but they require a firm understanding of the nature of the equations. Ignoring the corrections can significantly bias the results of experiments, if the magnitude of the individual data points varies significantly. For data that are more closely grouped, as many size distributions are, it introduces less error to ignore the requirement for different weighting factors.

A good example of the nature of the error can be found in the area and perimeter values discussed in Chapter 5. In an automatically determined set of measurements, the uncertain pixels are on the edge of the features, and so the error in the area is roughly proportional to the feature's perimeter. Different features with the same area and different shapes (different perimeters) would have different expected errors in the area values, and this should ideally be reflected in any mathematical fitting operation. However, this is by no means always done.

Chapter 2

Image Types

Our principal interest in applying the kinds of measurements that stereometry provides will be related to microscopic images of various kinds. The same methods apply in principle to macroscopic or even astronomical problems (for instance, counting the chips in Toll House cookies or estimating the number of beans in a jar, or the probability of striking an asteroid while passing from Mars to Jupiter), but it is with microscopic images that the greatest development of these methods has taken place, and we will leave it to the reader to substitute the proper terminology to adapt the arguments and descriptions to other fields.

However, even within the field of microscopy, there is an enormous range of possible image and sample types. First, there is the fundamental difference between the ideal random planar slice through a solid matrix, for instance a metal alloy containing grains and dispersed precipitate particles, versus other types of commonly encountered images. A very common and rather important field of measurements has to do with discrete particles, for instance powdered raw materials from ceramic processing or pharmaceutical production, dispersed on a flat substrate. This gives rise to a different class of measurements, as we shall see. A third type of image is that obtained by viewing in transmission through a section of finite thickness, which is commonly the case for the electron or light microscope, especially when applied to biological tissues. Again, this case calls for special models to relate what we see to what is there.

Planar Sections

Figure 1 shows a representative ideal section through a 3-D structure. This is almost always shown as a plane, because it makes the mathematical analysis simpler and corresponds in most cases to the way actual sections are prepared (by cutting, grinding, etc.).

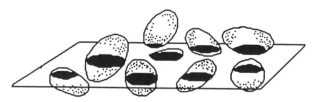

Figure 1. Schematic diagram of a planar section through material containing a dispersed phase. The intersections of the plane with the 3-D features produce features on the viewing plane which can be measured and interpreted.

There are some measurements that depend on the surface being a plane (for instance, determination of the mean curvature of surfaces), but most of the common measurements (volume fraction, number of features per unit volume, amount of surface area per unit volume, and so on) do not require this restriction. If your polished surface is not quite a perfect plane, it probably won't affect your results in any case, as the depth of field of the light microscope is so small that you will lose the ability to focus the image before the departure from a perfect plane makes any difference to the measurements.

This may not be the case with some other microscopes, for instance the Scanning Electron Microscope (SEM). This has great depth of field, and for that reason is often used to study quite rough or irregular surfaces. One particular type of specimen which it might be quite useful to study is a fracture surface in material. If it were practical to determine the area fraction of different fracture modes, or the number of dimples per unit area, or the total area of steps across cleavage surfaces, the results might correlate with measured fracture behavior.

One problem with looking at fracture surfaces in materials is that they may not represent valid "random" samples of the material structure. Presumably the fracture followed its particular path because of some weakness in the matrix, which may be structurally related (e.g. to too low or high a concentration of second phase particles, or grain size). Comparison of the structures seen on a fracture surface to those in a true random section may thus be interesting in their own right, but this is not easy to do.

Fracturing of frozen biological samples is also used as a preparative method, often specifically to expose a surface along a membrane where counting and measurement of features can take place. This is valid insofar as it represents the structure at or near the membrane, but the topography can still present measurement difficulties. Another biological example would be to determine the spacing between features such as cilia on the surface of an organism; it would be necessary to compensate for the fact that the surface is viewed at various angles over its irregular shape.

This is not simple to do. The roughness of the surface means that areas are foreshortened in the image, and the surface area is often considerably more than the projected area of the image. In fact, the ratio of the actual area to the planar area (the so-called "wrinkle factor") is one of the parameters which we would like to know, and it takes a considerable amount of work to get it. This involves stereoscopy rather than stereometry, that is, the use of two views at different angles to show the vertical relief of the surface as parallax in the two images. We will return to this subject in the final chapter, and for the time being will agree to restrict ourselves to reasonably flat (planar) surfaces for our sections.

From the drawing in Figure 1, it should be evident that sections cut through the structures within the 3-D specimen at random. Whether we are dealing with grains in a metal or mitochondria in cells, the section area is not likely to be as large as the maximum dimension of the feature. This means that we shall have to apply mathematical corrections to deduce their true size distribution. On the other hand, we may reasonably expect that if the features are randomly oriented in the 3-D specimen, or the random planes are placed at a variety of angles, we can sample (in the statistical

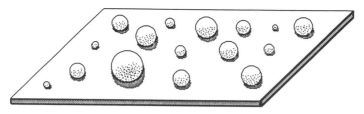

Figure 2. Schematic diagram of simple convex features
ideally dispersed on a substrate.

sense) all of the important dimensions of the features. If the structure is not random,
but has preferred orientation with respect to the outside world, we will see in in a later
chapter that there are ways to describe that, too.

Projected Images

Projected images give some very different types of information. Figure 2
shows an example, in which dispersed particles are sitting on a substrate (for instance
for viewing in the SEM or with reflected light in a light microscope).

Figure 3 indicates that a similar projected image may be obtained from a view
through a thick but transparent section, as in the transmission electron microscope
(TEM) or transmitted light microscope. In many cases the view through a reasonably
thick section (much thicker than the heights of any of the objects) may be similar in its
interpretation to that for the "projected image" case, but we will discuss the problems
of sections with finite thickness more fully shortly. In the case of the projected image,

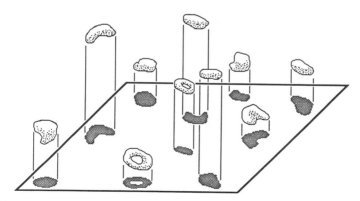

Figure 3. Schematic diagram of dispersed particles in a transparent medium,
with their projected images. Note that these images disguise the irregular
shape of the particles, particularly those which are concave.

we do see a maximum dimension. It is really a sort of "caliper" dimension, since the object may not be as wide at any one place as its shadow is. The shadow or caliper dimension is often referred to as a Feret's diameter, and is defined as the projected length of a feature in some predetermined direction. For particles on a substrate, it is important to keep in mind that the projected image has this nature and will often hide irregularities in shape, especially concavities on the feature surface.

Another potential problem with this type of image (or more properly with this type of sample) is the implicit assumption that we must ordinarily make that the vertical dimension of the features can be estimated from the projected image itself. The most common assumption is that the height of the particles is the same as the breadth of their projected shadows. For instance, if the particles were all footballs, we would expect to find them lying on their side rather than standing on end, and this would then correspond to the assumption above. Unfortunately, real particles are rarely as simple in shape as footballs, and furthermore microscopic ones have such low mass that it is not always likely that they will "fall over" onto their sides (other forces, including electrostatic ones, may cause them to line up in unexpected ways).

In some cases it may be permissible to assume that the features are randomly oriented, so that the height can be estimated as a mean diameter of the shadow. This is often the case with extraction replicas of metals for use in the TEM, where a plastic or carbon film is applied to a surface and then chemical etching is used to dissolve the matrix and allow particles to adhere to or collect on the film, which is removed for viewing (the thickness of the "section" is determined by the depth of chemical etching). In some other situations, independent knowledge of thicknesses may be available (for instance from the sample preparation method). But in many cases it may not be possible to completely avoid an arbitrary assumption about the thickness.

The figure shown is ideal in one other respect - the particles are all nicely separated. If they touched or overlapped, it might not be possible to make useful measurements on their size. If touching surfaces were tangential only, and the features were close enough to the same size and shape that there was no way for one to "hide" under another, then assumptions about convexity might permit the touching features to be separated arbitrarily. But in real samples with a wide range of sizes and shapes, most cases of incomplete separation create substantial bias in the results for

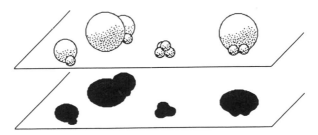

Figure 4. Example of touching and overlapping particles on a substrate, with their projected images

size distributions (usually in the direction of reporting too few small features, and too many large ones, or features that are too large). Figure 4 illustrates the problem: The shadow images cannot be reliably interpreted to determine the size the of features. If the features' shapes are not known, even their number may be indeterminate.

The solution to this problem lies primarily in the preparation of the specimens. If the natural material is well dispersed, there may be no problem. If the particles are sparse, like asbestos fibers in water, the practical difficulty may lie in finding enough features to measure in order to obtain adequate counting statistics. But a common situation is to have a container of a powder presented for measurement, by first dispersing it onto a suitable substrate (for instance a glass slide or an SEM stub). Techniques that are useful include entraining the particles in a gas or liquid (for instance using a perfume atomizer) and "spraying" them onto the substrate, which may have been coated with a thin layer of some adhesive first, or spreading the particles on a liquid where they will float because of surface tension (but not agglomerate into floating islands – sonic vibrations may be a help here), and then pick them up onto the substrate by dipping it into and removing it from the liquid.

A more heroic method which is sometimes used for difficult samples (Thaulow & White 71) requires making a mixture of camphor and napthalene (60:40% by weight). This composition corresponds to the eutectic. The material is a solid at room temperature and a liquid at slightly higher temperatures (the melting point is 32° C) A small amount of particulate material (a few percent by volume) can be added to this mixture and mechanically worked (one way is to seal the mixture in a plastic bag and knead it by hand) until the individual particles are well separated. Depending on the material, this may be easy or difficult. Then the mixture is spread (much like butter, which has about the same consistency) onto the substrate (typically a glass cover slide), and placed in a vacuum evaporator. The camphor-napthalene mixture has a relatively high vapor pressure, and will sublime; since it is a eutectic, there is no residual and the sublimation does not cause any rearrangement of the suspended particles (nor any displacement or agglomeration of the suspended particles during solidification after working the mixture). The result is the gradual settling of the particles onto the substrate (and, sometimes, the clogging of the vacuum system with the sublimed chemicals – use a good cold trap).

It is not the point of this discussion to present a comprehensive set of techniques that will be useful in all circumstances. Rather, it is fair to say that no such cookbook exists or is likely to. The variety of particulate materials that are of potential interest is vast, and includes many industrial raw materials and final products with electrostatic, magnetic (consider the oxide particles used to coat recording tape and computer disks, for which the size distribution is indeed critical) adhesive or other surface-active properties which make dispersal particularly difficult. Some creative work may be needed to obtain good dispersal. On the other hand, if clumps of particles exist in the final product, then it may be the size of the clumps that is important and must be measured by the stereometric measurement method (consider the pigments and clay extenders in many types of paint, for instance, which fall into this category). Finally, it is sometimes better to embed the recalcitrant agglomerates in a foreign matrix, and cut planar sections through the solid to reveal the particles, from which the kinds of measurements already mentioned can be made.

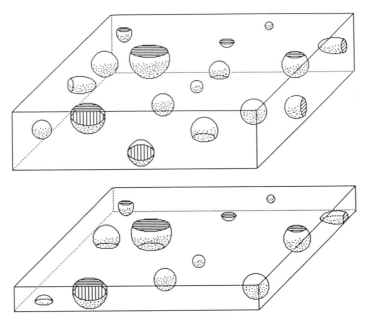

Figure 5. Sections through material containing a dispersed phase. The thicker section contains more features, which are hence more likely to overlap in the projected image. The thinner section is more likely to cut off portions of features so that the maximum dimensions are not revealed.

Finite Sections

Sections of finite thickness have the problems of both planar sections and projected images. The particles may overlap or touch, and in addition (as shown in Figure 5) they may be cut off so that the maximum dimension is not seen. The latter problem can be corrected, as we will see in a subsequent chapter, if the thickness of the section is known.

The thickness of sections can be determined in a variety of ways, including calibration of the device used to produce them (e.g. a microtome). More commonly, measurements are made on the sections after preparation. These may include weighing them, judging the color of light diffracted by them, measuring the absorbance of light or scattering of electrons in regions of embedding material surrounding the sample, or (for materials specimens in the electron microscope) measurement of diffraction effects. Generation of X-rays or calibration of the backscattered electron signal may also be used in the EM. One of the most common methods is using parallax: marks on the top and bottom surfaces (including the contamination marks that may develop in the electron microscope, or scratches visible in the light microscope) are shifted relative to each other by tilting the sample by a known amount, and from the shift, the thickness is calculated. This is identical in principle to stereoscopy, which will be discussed in a later chapter.

Space-filling structures and dispersed features

There are two principal classes of "features" (3-D objects to be measured) which we will encounter. They are dispersed, isolated features and continuous, space-filling structures. Examples of the former include precipitates in metals, bubbles from radiation damage in fuel rods, organelles within cells, and so on. Notice that bubbles, presumably filled with nothing, qualify as features. The criterion here is that the features are distinguishable from the surrounding matrix (and from each other, if they touch). The classic space-filling structure is that of grains in a metal. The boundaries between grains are visible in the images, but the grains are otherwise indistinguishable. By convention the boundaries are considered to be of zero thickness (on the scale of the image), although for cell membranes and similar structures this may not be true, and even for metals, the etching process used to reveal the boundaries may cause considerable broadening.

It is possible to have both kinds of features in the same sample (for instance a ceramic consisting of many grains of one identical material, with dispersed grains of a second phase, or voids). There are some kinds of measurements which apply to one class of features only, such as the volume fraction of the second phase. In the case of space-filling grains, parameters such as size, surface area per unit volume, and so on, are fairly simple to define and measure. In multiphase materials it may be appropriate to measure only a subset of some parameters; for instance, we might want to know the amount of surface area per unit volume separating phase A from phase B, but not involving interfaces of A with A, B with B, or either with C (this would be important in any situation in which reaction or diffusion across such an interface was important, and could equally apply to biological structures). In all these cases, we must presume that there is a way to distinguish the various phases from each other.

There is also the problem of the measurement "universe." Do we want to know the volume fraction of phase B in the total sample, or, if phase B is present as precipitates within phase A, do we want the measurements to apply only to that portion of all the B, and to be expressed as a fraction of phase A's volume (biologists: consider measuring vacuoles as a fraction of only one kind of cell, and ignoring intracellular spaces that are empty but are recognizably not vacuoles). In this case, it makes sense to initially restrict our measurements to the regions of our images which are phase A and ignore the rest. The use of reference areas that are not simply the rectangular or circular outline of the typical image is fairly easy to do, but we must be aware of the change, and the opportunity to introduce preferred orientation or other bias.

Usually we will consider the entire image area (typically square or rectangular) being measured as the reference space, so that if there are 10 objects visible in the field, covering 15% of it, and the total image area is 100 square microns, then these values will be used to determine the number of features per unit volume (Chapter 4)

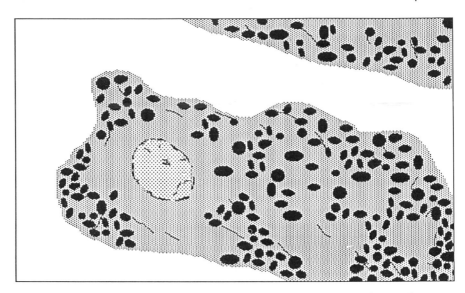

Figure 6. Simplified rendering of transmission electron microscope image of stained organelles within cells. Compare to Problem 5 in Appendix B, showing lysosomes in rat pancreas.

and the volume fraction (Chapter 3) of the objects. This is not always appropriate, however; it is sometimes useful to treat some other region as the reference area.

The most frequent example is that of cells seen in thin sections (in either the light or electron microscope). As shown in the schematic image of Figure 6, to determine the volume fraction of the organelles (dark features within the cells), and their density in the cells, it is necessary to determine not only the number and size of the features themselves, but also the fraction of the image area that the cells occupy (ignoring the white intercellular regions). This requires, in principle, two counting or measurement operations, one for the cells (or the white space) and another for the particles. These can sometimes be accomplished simultaneously, but it is important to keep in mind that the statistical precision for each value may be different, and that these will combine to limit the precision of the final result.

Another "universe" problem concerns the finite extent of our images, and their finite resolution. At low magnification, small features may be invisible (either too small to see or with too little contrast). At high magnifications, large features may extend beyond the boundaries of a single image frame and be impossible to measure. The solutions are 1) to work at several magnifications and combine the data (which must be done in proportion to the actual area viewed at each magnification, rather than in terms of the number of image frames or features counted); 2) to apply mathematical corrections to estimate by extrapolation the amount of small features (which works only for some kinds of size distributions); and/or 3) to apply corrections to the size distribution that is measured, to account for the likelihood of larger features intersecting the edge of the image frame (either mathematically, or by incorporating "guard frames" in the image). All of these techniques will be described where appropriate.

Types of images and contrast mechanisms

So far, the microscopist reading these discussions will probably have pictured mental images of the kinds most familiar in his or her routine work. These may be light or electron microscope images, using transmission (light or electrons) or surface (reflected light or secondary electrons) modes of operation. The contrast mechanisms are very different in these cases, and may influence what can be seen and measured. For instance, the simple need to distinguish between different structures (phases, organelles, etc.) for individual measurement and counting may be accomplished by chemical staining to introduce contrast due to the scattering of light or electrons, or by polarization of light (common for crystalline materials such as minerals), or in other ways. It is usually necessary to assume that the contrast is unequivocal, that is, that all of a certain class of features are revealed equally, and that everything that is stained, colored, or otherwise demarcated by a particular procedure belongs to the class of things of interest.

Of course, this is not always the case. One very common example in electron microscopy of materials has to do with dislocation counting (to determine the density of dislocations, which is related to the strain in the material and perhaps to its mechanical properties). Dislocations show up as dark lines in the transmitted electron image because the atomic lattice in the immediate vicinity of the dislocation is distorted, and this changes the electron diffraction there. This deflects electrons away from the image and produces the dark line. However, for any given orientation of the specimen, only some dislocations will show contrast (depending on their orientation with respect to the lattice, and the type of atomic offset in the dislocation). Furthermore, when one grain is oriented to show dislocations, others in the same image may not be. The consequence is that the image shows only a fraction of the dislocations present.

Sometimes this is as we wish (for instance, to study the dislocation density on a particular lattice plane). But generally it is necessary to make some assumptions about the relative abundance of the unseen dislocations to those which are imaged, perhaps by looking at images obtained under different diffraction conditions, and than estimate the "true" density from our partial measurements. Note the opportunity for statistical degradation of the answer in this situation, either due to bias or simply magnification of normal counting error due to the multiplication operation (propagation of errors).

There are other similar cases that arise when using polarized light, backscattered electrons, and so on. One common solution is to obtain several images using different contrast mechanisms, and then combine then to isolate all of the features of interest. For example, consider the backscattered electron image in the SEM, which has a brightness proportional to the mean atomic number of the target (in the absence of surface irregularities). This may allow the isolation of one phase from others, but there may be several phases with very similar average atomic numbers that cannot be separated in this way.

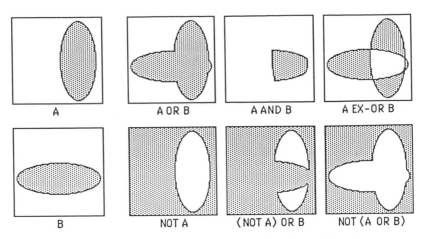

Figure 7. Two representative images showing regions A and B, and several of the possible combinations possible using the *OR*, *AND*, *EXCLUSIVE–OR* and *NOT* operators.

On the other hand, a second image (perhaps using characteristic X-rays from iron, which would isolate all of the iron-containing phases, or the secondary electron image, which would isolate phases with a particular surface work function) may be able to select the phase of interest plus some different confusing ones. Then logical combination of the images, in this example using an *AND* function, can provide just the phase of interest for measurement and counting. Figure 7 shows examples of the use of the logical *AND*, *OR* and *EXCLUSIVE–OR*. For each, the logical *NOT* can also be used, for instance by using a photographic negative instead of a positive. This technique is readily carried out using darkroom techniques to superimpose multiple images, and is also common in many computer-based image measuring systems which can store the images internally and perform the logical combinations in memory.

In mathematical terms, the *OR* function combines two sets of points (points lying within feature outlines as defined by two different selection criteria) by producing a set in which each point is part of the combination if either of the two original sets contained it. The *AND* function produces a combined image with those points which are present in both original sets (but not those which are present in only one or the other). The *EXCLUSIVE–OR* function selects those points which are present in one original set or the other, but not both. The *NOT* function simply selects all those points which are outside the original set. All of these can be combined in any sequence, but is important to note (as shown graphically in the figure) that some of them are not commutative (ie. it makes a difference which order they are written in, and parentheses are important!)

Sampling

Finally, there may be (indeed, there all too often are) cases in which no unequivocal separation of the features of interest from the remainder of the image can be achieved using these "simple" methods (i.e. in principle totally free from human judgement, and hence reproducible). It may be that dispersed particles can only be separated into two classes, one we want to count and the other we do not, based on size or shape. If the features are measured, then we can ignore the ones that do not match the limits set. But it is a human decision to skip some and accept others, and if that judgement is wrong, then the results will be biased in a way that is hidden from the person who reads the final report.

Particularly in biological tissue, where the contrast mechanisms are complex and the structures vary greatly in appearance, there are many instances where no unequivocal demarcation of features can be made based on contrast, or even shape, and it falls to an "experienced" operator to decide which are the structures of interest. It may not even be easy to describe the criteria that are used. The power of the human "image processing and recognition" computer is vast, and is often able to distinguish very subtle details to permit the required identification. However, this capability is also notorious for being difficult to transmit from one operator to another, and may vary from one time of day (or week) to another. Extensive statistical comparisons of data taken on duplicate images, or multiple images from duplicate samples, are vital to assess the bias that may be present between operators or different laboratories in these cases.

As mentioned in Chapter 1, there are situations in which random sampling is inappropriate. If the specimen is grossly heterogeneous in a known way, then specific samples must be taken in an orderly fashion to represent the entire material. For example, the roots, stems, branches and leaves of a tree must be sampled separately to learn about the tree. The same approach is often used with animals (all or part) and other organic specimens, but would be equally appropriate to distinguish the internal and near-surface structure of a cast metal, for example. The key to this approach is to logically subdivide the entire specimen into a series of smaller ones, which can each be presumed to be homogeneous and therefore is a legitimate candidate for random sampling. There is also the possibility of comprehensive sampling, where a series of regularly spaced and planar sections are used to fully cover a specimen, and an orderly array of image locations is examined on each section.

As a final caution, random sampling means just that. It is essential to position the specimen for measurement, counting, etc. without looking at the image. Subsequent desire to "touch up" the position to make a prettier picture is strictly forbidden, and will inevitably introduce bias. The over-used phrase "typical microstructural image" usually means, in reality, that the published picture is the best or prettiest taken, and certainly not a typical "random" one.

For most purposes, measurements can be made equally well using photographic prints, or negatives, or working with a live image from a microscope (and performing measurements or counts on a viewing screen or with an eyepiece reticle). There are some relative advantages and disadvantages to each approach.

Photographic prints can control image contrast to make it easier to distinguish features and boundaries, but at the same time, areas that are black in the print will tend to grow slightly in the printing process, which can introduce bias. Prints have the advantage that they can be marked as an aid to counting, and of course the same image can be measured again if the results must later be checked. But where it is appropriate, working with the image directly is much faster and may allow more measurements to be made (for correspondingly better statistics).

Chapter 3

Manual Methods

Volume Fraction

The oldest stereometric technique, and perhaps the most intuitively obvious and hence easy to understand, is the determination of volume fraction of objects in a structure or of a phase in a material. We will use it here as an introduction to the various types of measurements. On a random section, as shown in Figure 1, different objects or phases can be discriminated, for instance by brightness or color in some suitable image formed with electrons or photons. More than 100 years ago, long before there was a field known as "stereology" (the word was coined by Hans Elias in 1961, at the first meeting of the interdisciplinary group of scientists which has evolved into an international society with periodic meetings), a French geologist (Delesse 1848) showed that the volume fraction of the phase was equal to the area fraction of the intersections of the planar section with the phase. (We shall use the word "phase" here to describe a recognizable structure within the solid; it could as easily be the nucleus within a cell, the bones within a vertebrate, or oil-containing shale within a mineral formation. The student is asked to mentally substitute the terminology appropriate to his or her interests.)

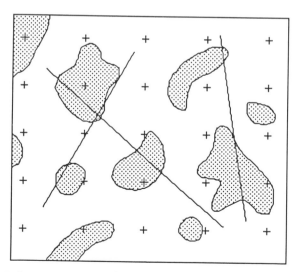

Figure 1. A representation of a section through a solid, with regions of one type of object or phase (shaded) dispersed in a matrix. The points and lines are added to facilitate the various measurements that will be described.

In what has become fairly standard nomenclature in stereology, we would write this simple truth as

$$V_V = A_A$$

This is read as "the volume per unit volume (or volume fraction) of the phase or object is equal to the area per unit area (or area fraction)". In addition to V and A, the letters L (lines), P (points), and N (number of something counted) will be used. Hence, P_L would be the number of points per unit line length, or N_A would be the number of features per unit area. The usage will become clear as more examples are shown.

In other words, it is possible to determine the volume fraction of the phase by measuring the areas of the intersections, or profiles outlining the phase in the plane section. Hence, the fundamental goal of stereology, which is to be able to infer 3-dimensional information from things seen on a two-dimensional surface, is indeed possible. The derivation of this principle can be found in several of the books cited in the bibliography. For our needs here, it is only necessary to note that the truth of the relationship depends only on the "randomness" of the plane with respect to the structure of the material, and not at all on the way in which the phase or object is shaped, dispersed or organized in the matrix. And, as always, we must note that there is the statistical problem of sampling; it may require looking at a number of different sections to determine the mean volume fraction and the standard deviation.

Of course, it is not really all that convenient to measure the areas on the individual sections. One way, actually used in many laboratories over the years, is to carefully cut out the phase regions or objects from photomicrographs of the sample, and then weigh them. The weight fraction of the photo areas from the phase, with respect to the total weight of the picture (and assuming uniform density photo paper) gives the area fraction and hence the volume fraction. Bias due to a human tendency to cut around the outline, rather than on it, is only one of the difficulties with this method.

About 50 years after Delesse, a further simplification was made. Another geologist, Rosiwal (1898) showed that another simple fraction, the fractional linear intercept, was also equal to the area and volume fractions. In our nomenclature

$$V_V = A_A = L_L$$

The fractional linear intercept is obtained by drawing many (random) lines on the image and measuring the length that is within the phase to be measured, divided by the total line length. Figure 1 shows some such lines. Measuring the length of each line and the length of the portion that lies within the shaded phase is easier to do than cutting and weighing the areas, and is at the heart of a device called the Hurlbut Counter (39), which was a common accessory in the metallography labs of one generation ago. The device consisted of a motor drive that turned one axis of motion of the microscope stage, to which was attached a turns counter. The operator looked through the microscope, which was fitted with a crosshair reticle, and visually identified the phase which was passing under the crosshair at any moment. For each phase, there was an assigned button on the counter mechanism that engaged another turns counter. After a suitable amount of sample had been traversed, the master counter indicated the total line length and the individual phase counters gave (after simple division) the fractional linear intercept, and hence the volume fraction of each phase. Electronic adaptations of the same idea to the SEM and electron microprobe have also been made (Dorfler 68).

In the 1930's and 40's a further improvement in technique was made possible by the demonstration of a number of investigators that yet another simple fraction could be used (Thomson 30). This is the point count fraction, determined by laying down a "random" distribution of points on the section image and counting the number which happen to hit the phase of interest. Counting has already been mentioned as lying at the heart of many stereological techniques, because it is relatively fast, has well understood errors, and in many cases can be automated. This is a good example. The location of the points to be counted is often taken as the intersections in a grid, as shown in Figure 1.

This pattern could be made into a transparent overlay to be placed on micrographs, or built into an eyepiece reticle to fit directly into the microscope (or marked on the CRT display of an electron microscope). Other similar patterns are available in eyepiece reticles for many light microscopes, to facilitate counting operations directly while viewing the sample, and we will see some of them shortly. It might seem that such a regular array of points could hardly qualify as "random," but indeed they are, if the structure itself is the randomizing agent.

The points should be spaced apart far enough that it is rare for two of them to fall into the same profile of the measured phase. This is simply to assure that the expected statistical precision is obtained, in terms of the number of points counted. Many automatic systems (which are discussed in a subsequent chapter) count 50,000 or more points per image, but the result is not any better than would be obtained from many fewer points because they are largely redundant. This is acceptable for the automatic systems, because no time would be saved by counting fewer points. However, when manual counting is used, a grid with from 25 to 100 points is most commonly used, and the image magnification should be adjusted so that the structure of the material is efficiently sampled. It is most efficient to count points in the minor phase, and to determine the major phase by difference.

Another problem to recognize is that while the phase or object boundaries may ideally be considered to have zero thickness (if the boundaries were actually broad, we would probably want to consider them to be another structure, as in the case of membranes), in the typical microscope image they have been broadened by etching, etc. to improve their visibility. Hence it will sometimes happen that a test point appears to lie on the boundary rather than inside or outside the phase. The usual rule in that case is to count it as 1/2. In Figure 1, the point fraction using the marked grid is 5.5 hits out of 25 points, giving an estimated volume fraction of 22% (of course, many more points would have to be counted to achieve reasonable precision in the result). The same consideration extends to the finite size of the test points or grid lines themselves.

For sections of finite thickness, the projected image will show (on the average) too great an area fraction for the one phase or set of objects, which is opaque. The same problem can arise due to etching or polishing relief in metals, where one phase is harder than the other (Figure 2), or in freeze-fractured and etched

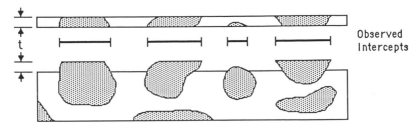

Figure 2. Observed intercept areas and lengths from a section of finite thickness, or with polishing relief, are greater than the true planar intercepts.

biological materials. Cahn and Nutting (59) have shown that the corrected area fraction can be obtained as

$$A_A{}' = A_A + S_V \cdot t / 4$$

where t is the thickness of the section, or the relief, and S_V is the surface density which will be described next.

We would now write the complete relationship as

$$V_V = A_A = L_L = P_P$$

where in each case, the nomenclature indicates (for volumes V, areas A, line lengths L or point counts P) a fraction of the total amount. Notice that the dimensionality of these expressions is all the same. It is often very important to pay attention to the dimensionality of stereometric relationships. The volume fraction (and the other things equal to it) has no units, and is given as a fraction or percent. Many of the other properties that we measure will have units, for instance of area or inverse length. For these, it will be important to know the image magnification; but for volume fraction, it is not needed.

When the reference area is not the entire image area, then this must be included in the calculation. For instance, for the stained (dark) lysosomes in Figure 6 of Chapter 2, the volume fraction within the cell is determined by the area fraction (or point count fraction, etc.) within the cells. If the point count for cells in the total image frame is P_{Pcell}, and the point count for lysosomes in the total image frame is P_{Plys}, then the volume fraction of lysosomes within the cells is P_{Pcell} times P_{Plys}. Similar relationships hold, of course, for L_L and A_A.

Surface Density

Another fundamental property that can be readily determined is the surface density, or the amount of surface area of objects or a phase contained in volume V of the solid. In our nomenclature, this is written as S_V. (S is often used to denote surfaces, indicating the possibility that they are curved, while A is commonly reserved for planar sections.) It may be interesting because of diffusion across the surface, free energy considerations, or other reasons. It applies equally well to many biological specimens (which are rich in membranes whose area can be determined in this way), as well as engineering materials (grain boundaries in metals, ceramics, composites) and so on.

The surface density problem has a fascinating history, which also sheds some light on the overall history of stereology. It can be linked in underlying mathematical principles to a problem solved in 1777 by Buffon, and known as the Buffon Needle problem. This is an exercise in geometric probability, which can be done analytically, but can also be verified using a computer with a "Monte-Carlo" technique (see Appendix A on geometric probability). This approach will be important later in testing other more complex measurement models, and this is a good place to develop the familiarity with the basic skills. Any computer and language, provided the system has a useful generator of psuedo-random numbers available, will suffice.

The Buffon problem is as follows: Given a surface like the striped portion of the American flag, in which the width of the stripes is d and the boundary width is negligible, and given a needle of length L, which is to be dropped at random onto the surface, what is the probability that the needle will intersect a boundary? The analytical answer is obtained by integrating over all displacements and rotations which the needle can assume, and turns out to be (as Buffon determined): $2L / \pi d$.

The Monte-Carlo approach uses random numbers to select positions and orientations for the needle, and counts the fraction of times the needle intersects a boundary. Given enough statistics (enough independent trials of placing the needle), the answer approaches the analytical value. This may seem redundant in this case, because we can solve it more efficiently with the analytical approach. But there are many other geometric probability problems which we can encounter in stereology which are not so easy to solve analytically, but for which the Monte-Carlo method is still fairly easy to program into a digital computer, even a very small one. Given enough computer running time (enough trials), useful results will be obtained.

But now back to our immediate problem, which is to determine the surface area per unit volume represented by boundaries on the planar section we see in our microscope image. Based simply on dimensionality, this will clearly have units of length squared (surface area) divided by length cubed (volume), or $(length)^{-1}$. Candidates for measurement would then be length per unit area (the length of the boundary lines divided by the area of the image) or points per unit length. It turns out that the most efficient way to obtain an answer is to draw random test lines of total length L and count the number of intersections of the lines with the boundaries or interest, P. This is shown schematically in Figure 3.

Drawing the lines and counting the intersection points shown in the figure gives 14.5 intersections. By convention, we count 1/2 for a tangent point. Also, one line has an end point within the shaded phase, so there are an odd number of intersections on that line. Then the surface area per unit volume is

$$S_V = 2\,P_L$$

or two times the number of points divided by the length of the line. The dimensions are right (*area/volume = 1/length*), but where did the factor of two come from? The answer is closely related to the Buffon needle problem, and can be obtained in the same way, by integrating over all the ways the test line can intersect the boundary

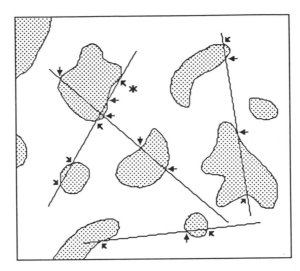

Figure 3. Random lines placed on a microstructure to count the number of intersections they make with the boundaries of the shaded objects or phase (arrows). At the point marked with the asterisk, the line is tangent to the boundary.

(hint: the angle will not in general be 90 degrees, so parallel test lines would produce intersections along the boundary separated by a greater distance than their spacing).

Two somewhat different derivations are given in the books by Underwood and Weibel, and in fact the derivation of this apparently simple result has been discovered and published quite independently at least eight times since 1945 (Saltykov), illustrating both the importance of the relationship and the wide diversity of fields in which stereology can play an important role (and the general lack of communication between them).

Note: an alternative way to arrive at the same measurement is to use the relationship from the Buffon problem, namely that the length of boundaries on the section area can be related to the number of intersections with the test lines as $B_A = P_L \pi/2$, but since $P_L = S_V/2$, we could obtain $S_V = 4B_A/\pi$. (B_A is the length of boundary line per unit area, S_V is the amount of surface per unit volume of the same boundary. The ratio $4/\pi$ enters because the boundaries do not intersect the plane section surface at a right angle). This relationship would be useful if we had a measurement of the length of the boundary line, which some semi-automatic and automatic machines will measure as the length of perimeter around the phase or objects of interest.

A word of caution: There are some measurements that are closely related, or even equivalent to the S_V parameter as it has just been defined here, but which may appear confusing because of the way terms are defined. For instance, for dispersed convex particles in a matrix, the surface area per unit volume may be given as four (not two) times the number of intersections of the test lines with particles, per unit length of line. The difference is simply resolved when you realize that for convex particles, the number of boundary intersections will be just twice the number of intersections with particles, so the two values are equivalent. If we know the volume fraction for the phase (which can be determined simultaneously from L_L with the same lines as used for the intersection count, or by using many lines and counting P_P using the end points of the line segments as points, or by using a grid and counting P_P from the points where the grid lines cross each other and P_L from intersections of the grid lines with the boundaries on the image), we can then get the actual surface area per unit volume of the dispersed phase.

Contiguity

An interesting parameter which may be determined directly from the surface area per unit volume is the degree of contiguity between phases. For instance, as shown in figure 4, a complex multiphase material may have places where a region of phase α touches a region of phase β, or phase γ, or even another distinguishable grain of phase α. If the arrangement of the phases is random within the structure, as might result from mixing of discrete particles, we would expect the chance of encountering each neighboring phase would simply be equal to the volume fraction of the phase. Variations from this expectation may indicate degrees of order in the material that have arisen in various physical ways (for instance, nuclei in cells always touch the cytoplasm, not other nuclei; grains of one composition in a complex ceramic are likely to cluster together because of sintering and diffusion reactions; magnetic fields may cause iron oxide particles to cluster in coatings).

The parameter that describes this sort of neighbor analysis is called contiguity. The contiguity of phases α and β in a multiphase material is simply given by the ratio of the surface area shared by the two phases, to the total surface area in the material or to all the surface area for one of the phases. We have already seen

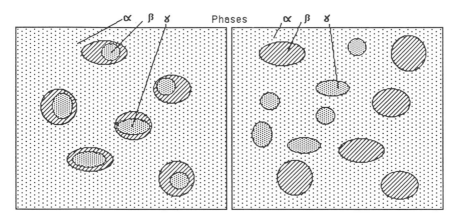

Figure 4. Two multiphase materials with extremes of contiguity between phases α, β and γ.

how to determine surface area by counting intercepts along random test lines. Now we must count two (or more) types of intercept: intercepts with boundaries between phase α and phase β, and all others. Then the contiguity of α with β is (Gurland 58)

$$S_{V\alpha\beta} / S_{Vtotal} = P_{L\alpha\beta} / P_{Ltotal}$$

Mean Intercept Length

There is another parameter that can be obtained from the intersection of test lines with objects. It is the mean intercept length, defined as the average length of the line along the chord that passes through the outline of the feature on the section. This is shown schematically in Figure 5.

There is only a single intercept along each line for convex particles, but there may be more than one intersection for concave outlines. The mean intercept length is conventionally written as L_3, where the subscript 3 acts as a reminder that we are dealing with an estimator of size for the three-dimensional body being sampled. It is determined by averaging over a number of test lines and may be written as L_L/N_L using our usual terminology. Using relationships given above, it has been shown (Tomkeieff 45) that

$$L_3 = 4 V_V / S_V \qquad \text{for a dispersed phase, and}$$

$$L_3 = 4 V / S \qquad \text{for a single particle.}$$

It also follows that the mean surface to volume ratio for particles cells, grains or other objects is

$$(S / V)_{mean} = 4 / L_{3\ mean}$$

Another parameter of considerable importance in materials is the mean free distance between particles (for instance, in a metal the distance between precipitates controls the pinning of dislocations and hence mechanical properties). This is written λ, and is the mean edge-to-edge uninterrupted distance between all possible pairs of

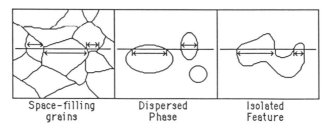

| Space-filling grains | Dispersed Phase | Isolated Feature |

Figure 5. Intercept lengths (marked with arrows) of a test line with different types of features: a) space-filling cells or grains in a metal; b) dispersed regions (particles or phases) in a matrix; c) isolated features.

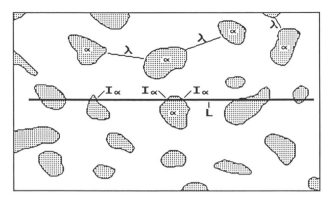

Figure 6. Dispersed phase in a matrix, showing the free spacing between particles of type α and the intercepts $l\alpha$ along a test line L, from which the mean value can be determined.

particles in the matrix (Fullman 53). As shown schematically in Figure 6, this can also be obtained from L_3 and the volume fraction of the phase, as

$$\lambda = L_{3\ mean}\ (1 - V_{V\alpha})\ /\ V_{V\alpha} = (1 - V_V)/N_L$$

It is important to note that this result is completely independent of the shape of the features or their size distribution. Either the mean free spacing, or the mean intercept length which is related directly to it, can therefore be used to correlate with mechanical properties in many systems. From the mean intercept length, we will see shortly that we can also determine the grain size of a space-filling phase, the size distribution of features of uniform (known) shape, and other parameters of considerable interest.

Note that the mean free distance λ is the average distance between all pairs of features in the matrix, and not the distance to the nearest neighbor. Chandrashekhar (43) has shown that for a distribution of points in space (he was concerned with stars in a galaxy) the mean nearest neighbor distance can be estimated as

$$\Delta_3 \cong 0.554\ P_V^{-1/3} = 0.554\ (\pi\ \rho^2\ \lambda)^{1/3}$$

With well dispersed, small size features having a low volume fraction, this should hold approximately for the distance between neighbors in solids as well. For small features on a plane, there is a similar relationship

$$\Delta_2 \cong 0.5\ N_A^{-1/2}$$

Line Density

If a line is suspended in a three-dimensional volume (for instance, dislocations in a metal, filaments in cells, plant roots in soils, or anything else that can be approximated as an ideal line, even if it has some finite lateral dimension), the

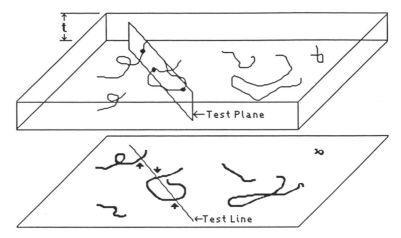

Figure 7. View of dislocations in a thin section, and their projected image. The test line on the image represents a test plane in the section, used to determine dislocation density.

length of the line per unit volume will have units of length^{-2}. On the surface being examined, the line intersections will appear as points, so we would only be able to count points per unit area P_A. Following somewhat the same reasoning as for the intersections of lines with boundaries, we can find that the factor two appears in the relationship for length density

$$L_V = 2\, P_A$$

The determination of linear density is frequently used with dislocations in metals. There are two main approaches taken: 1) counting of etch pits on a surface, where each pit marks the emergence of a dislocation and hence counting them gives P_A directly, and 2) viewing dislocations in a section of finite thickness in the TEM. In the latter case, which is shown schematically in Figure 6, it is possible to draw a line on the microscope image which represents a plane oriented vertically in the section.

The number of intersections of the dislocations with the line N_L is just the same as the number of intersections with the plane (P_A above). The area of the plane is L times t where L is the length of the line and t is the section thickness. Hence

$$L_V = 2\, N_L\, /\, t$$

Grain size determination

A parameter of great practical interest, especially in the metallurgical field (although it is used for other space-filling structures as well), and with considerable history as a stereometric parameter, particularly in the field of steel making, is the grain size number. Several different methods have been used to describe this, but they have been rationalized and made to give the same result by efforts of committee *E4* (Metallography) of *ASTM* (the American Society for Testing and Materials). The most recent revision of the method (standard *E112*) defines a grain size number *G* in terms of *n*, the number of grains per square inch at 100X magnification

$$n = 2^{(G-1)}$$

which is equivalent to

$$G = 1 + \log_{10} n / \log_{10} 2$$

Most countries which use a metric system have a similar relationship based on *m*, the number of grains per square millimeter at 1X magnification

$$G_m = (\log_{10} m / \log_{10} 2) - 2.95$$

which produces results that are identical within the fundamental precision of the method (results are sometimes written with a single digit to the right of the decimal point, but when it is used as a specification for metal structure, only numbers rounded to the nearest 0.5 are used).

There are three different measurement methods available to determine grain size, in addition to visual comparison to standard sizes in published charts. It will be instructive to see why (and to what extent) they are equivalent, and how they work.

The most direct application of the definition of the *ASTM* grain size number is the Jeffries (16), or planimetric method, which simply counts the number of grains per square inch at a known magnification (preferably 100X, but any other can be accommodated). The square inch is usually not square, but round to minimize problems with preferred orientation. Draw a circle of known area on the

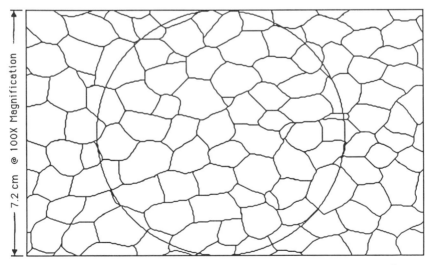

Figure 8. Grain counting method. The circle has an area of 0.4072 mm^2, and contains 37 grains and 30 "half grains" which intersect the edge of the test area. The result is a grain size calculation of 4.05 $\cong$ 4

photomicrograph, or directly on the viewing screen of the microscope. In the latter case, adjust the magnification so that there are at least 50 grains in the measurement area. Otherwise, several areas may need to be counted to obtain adequate precision. Count the grains within the test area, and for grains that intersect the perimeter, use a count of 1/2 regardless of the extent to which they lie inside or outside the area. Then use N the number of grains, and M the magnification, to calculate:

$$G = (\ 3.322 \log_{10} (NM^2/area)\ ) -2.95$$

(for multiple areas, adjust the value accordingly.) Figure 8 shows this method applied to the image of grain outlines shown in the first chapter. The result for the image area chosen is a grain size of 4.

The problem with the Jeffries method is that it usually requires marking the photograph to get an accurate count, and hence it is slow. A much faster result with equivalent precision can be obtained by making a count of intercepts with a test line. Either random lines can be drawn on the image, or circles can be used (the latter are often preferred because they avoid problems with preferred orientation of the grains). An overlay or eyepiece reticle with lines or circles of known length is often used. Counting of intercepts along the line(s) is easier than counting grains within the area, and does not usually require any marking to keep track of where you are. Plus, of course, other parameters that can be derived from the intercept length such as S_V and λ, are obtained simultaneously.

The intercept, or Heyn method (03) produces the result N_L (number of intercepts per unit length of test line), from which L_3 is determined as $1/N_L$, and G follows directly as

$$G = (-6.6457 \log_{10} L_3) - 3.298$$

for L_3 in millimeters. When line segments, rather than continuous circles are used to perform the count, it is conventional to add 1/2 for each line end when it lies within a grain (rather than exactly on a boundary). Also, tangent counts (where the line appears to touch a grain boundary but not cross it) are counted as 1/2, and triple point counts (where the test line appears to pass exactly through a branch point where two grain boundaries meet) are counted as 3/2. Notice that as promised, we have now related grain size to the mean intercept length. If we use the value of 12.54 intersections per mm given for the figure in Chapter 1, the result is a grain size of exactly 4.

A third equivalent method, admittedly the least used, is the Euler or triple-point method. The number of vertices of polygons is related to the number of polygons (grains), and so the number of triple-points per unit area can also be counted, instead of the grains themselves. If this is done, it is conventional to count a 4-rayed point as two triple points. This is because such structures do not really exist in metal grains (given enough magnification, they resolve into two nearby triple points). Then the number of grains per unit area N_A is determined as

$$N_A = (\ P/2 + 1\ ) / area$$

from which the grain size can be determined using the same expression given before.

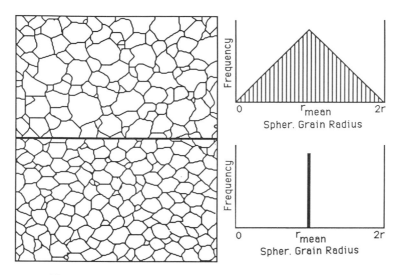

Figure 9. Microstructures with the same mean grain size. One has a uniform size for all grains, and the second has a distribution of sizes up to twice the mean, as shown.

The hidden assumption in these grain size measurements based on a mean intercept length or a mean number of grains per unit area, is that there is a mean grain size that really represents the structure. But not all metals have a uniform grain size; duplex and more complicated distributions of individual grain sizes are common, and the average result of the single grain size number may inadequately describe this.

Figure 9 shows two microstructures, each with the same mean grain size. However, one has a uniform grain size (the variation in the size of the grains seen on the test plane is due to the probability of intersecting the grains at different positions), while the second sample has a distribution of grain sizes up to twice the mean, as shown in the figure. There is a substantial difference in the appearance of the two microstructures The "grain size number" is not properly applicable to duplex grain structures. In the next chapter we will learn how to characterize distributions of feature or grain sizes.

The grain size numbers determined by the three different methods are the same only for the ideal case of uniform, equiaxed, space filling grains. If the material contains an intimate mixture of two (or more) sizes of grains, the intercept method will yield a lower grain size number (larger average grain size) than will the planimetric method. Both methods deal inexactly (and differently) with grains that are not equiaxed, or with structures that contain a second phase (e.g. inclusions in steels). The latter are usually ignored in counting grains, and may be ignored in counting intercepts, although they affect the mean free path and the mechanical properties which depend upon it. Since the intercept method is directly related to the mean free path, while the grain counting method is not, the former should be preferred in cases where the two do not agree.

Curvature

The curvature of surfaces in three dimensions can be described by two radii, that of the largest and smallest circles which can be placed tangent to the surface (as shown in Figure 10). When these are both positive, the surface is locally convex. If they are both negative, the surface is locally concave, and if their signs are opposite, the location is a saddle point. If one is infinite, the surface is cylindrical, and obviously both are infinite (zero curvature) for a locally flat surface. Two kinds of curvature are defined for surface, the mean curvature

$$K = 1/2 \ (1/r_1 + 1/r_2)$$

and the total or Gaussian curvature

$$G = 1 / (r_1 \cdot r_2)$$

Since the Gaussian curvature for any closed feature (regardless of shape) integrates to 4π, the total Gaussian curvature of the surface in any specimen is simply 4π times the number of particles, which we will learn how to determine in the next chapter. Minkowski (03) has shown that the mean curvature K for a convex shape is equal to $2\pi d$ where d is the mean caliper diameter (which is used in Chapter 4). This implies that the total curvature (either the mean curvature, for convex bodies, or the Gaussian curvature, for any shape bodies) in a section is related to the particle density in the structure. Chapter 4 will consider ways to determine N_V in more detail. The mean curvature is also proportional to size for some other particular types of non-convex objects, such as the length of tubules or fibers, and the perimeter of "muralia" (sheets or membranes) viewed in a section.

The mean curvature is a particularly useful parameter for many kinds or real interfaces. For instance, it is related to p, the surface pressure on an interface and γ, the surface tension coefficient of the boundary, as

$$p = 2 \ \gamma K_{mean}$$

For example, Gil and Weibel (72) have used this relationship as a tool to determine the pressure on alveolar surfaces by measuring the curvature of boundary profiles seen in cross sections through lungs. Many similar potential applications exist in biological systems (any situation where there is a pressure difference across a

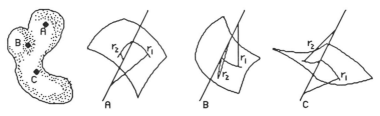

Figure 10. Surface of a complex shape, showing convex, concave and saddle points.

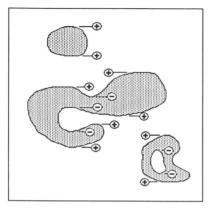

Tangent count
using horizontal
test direction
to determine
Mean Curvature
$M_V = \pi (T_+ - T_-)/A$

Figure 11. Representative microstructure for determining the integrated mean curvature from the net tangent count. Points marked + and − are positive and negative tangents with a horizontal test line.

membrane, such as cell membranes in root systems, etc.) as well as for materials problems (bubble formation, surface tension at phase boundaries, and so forth).

To determine the mean curvature in a specimen, when we cannot see the entire surfaces but only their intersection with one plane, may seem difficult, requiring measurement of radii of curvature and other things. Fortunately, this is not the case. Dehoff and Cahn simultaneously (67) and independently derived and published a simple method that only requires counting points. The points to be counted are the tangent points where the boundaries would meet a test line swept across the image with a single arbitrary orientation, which for practical reasons we usually choose to be parallel to one edge of the image.

Figure 11 shows the method schematically. The tangent points where the boundary is locally horizontal are marked on the image (any other orientation could also be used; we assume here that the structure is isotropic). We distinguish two kinds of tangent points, positive and negative. Positive tangent points are places where the local curvature is convex (the radius of a fitted circle would point into the feature), and vice versa. The integral mean curvature per unit volume is then

$$M_V = \pi (T_+ - T_-) / A$$

where T_+ and T_- are the total number of positive and negative tangent counts, and A is the area sampled. Notice that for the case of purely convex particles, there will be two T_+ counts and zero T_- counts for each, and the total mean curvature will be $2\pi N_A$ where N_A is the number of features per unit area.

Reticles to aid counting

Many of the techniques described in this chapter are routinely performed by human observers, often working directly at the microscope rather than with photographic prints. To facilitate the counting operations, a variety of special reticles

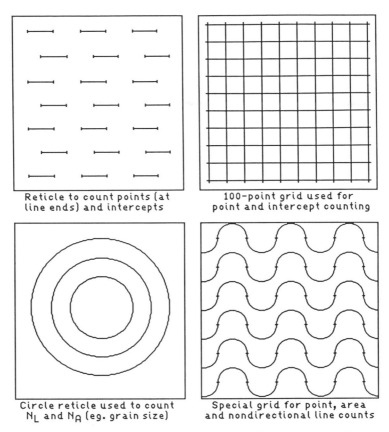

Reticle to count points (at line ends) and intercepts	100-point grid used for point and intercept counting
Circle reticle used to count N_L and N_A (eg. grain size)	Special grid for point, area and nondirectional line counts

Figure 12. Some representative eyepiece reticles used for counting.

have been devised that can be placed in the microscope eyepiece. In some cases, these are superimposed instead on a projection screen or a video display from a camera attached to the microscope, but the principle is the same. Some of the more common reticles are shown in Figure 12.

Probably the most common of all these is the simple grid with (typically) 100 points. This is used to determine volume fraction, by counting the number of points (at the intersections of the grid lines) that lie in a particular phase. The lines make it easier to see the intersections than if they were present just as points, and if boundary intersections with the grid lines are counted, S_V can also be determined as was described before.

Points for counting the volume fraction are also built into two of the other grids, but they incorporate other features as well. The reticle with the array of short lines (Weibel 66) can be used to count intercepts for surface area measurement, and since the volume fraction is determined from a point count using the line end points, the mean surface to volume ratio for the phase of interest can be determined directly.

The lines also help the eye to count horizontal tangent points, for the total mean curvature determination.

The grid with the "wavy" lines was developed by Merz (67) to minimize the sensitivity of reticles with straight lines to anisotropy in the specimen. Intersections can be counted with the semicircular lines, and the points define a grid for volume fraction determination. In addition, the points mark out square areas which may be used to determine individual P_A values, from which a mean and standard deviation can be calculated. For all these reticle types, the total line length, total area and total number of points are known beforehand, to facilitate computation.

This is also the case with the reticle having three concentric circles. This is particularly useful for grain size determinations. The outer circle has a radius of 79.8 mm. at a magnification of 100X, so that the area within the circle is just 5000 mm^2. This facilitates calculating grain size when the number of grains within the circle is counted. The total length of the three lines is just 500 mm (at 100X), so that the total number of intersections with grain boundaries can be quickly and easily converted to grain size as well. The use of circles minimizes anisotropic effects on the results.

Many other specialized reticles are also available, including ones with a series of circles, used to measure the size of features (as will be discussed in the next chapter). If enough of a particular type of work is to be done routinely, and it is practical to do it on the microscope without resorting to photographs, then a properly designed reticle will quickly pay for itself in time saved, and in improved precision.

Magnification and units

This chapter has presented the most commonly used stereometric relationships for commonly measured parameters using manual methods. There is a rich literature in theoretical stereology, and a wide variety of specific measurements have been devised to characterize particular types of samples or describe parameters that have meaning in a given instance. Some of the relationships are complex, and many of the derivations are further obfuscated by non-standard symbology. It will be of great assistance in trying to evaluate the usefulness and even the accuracy of these methods to keep in mind the dimensionality of various kinds of measurements.

The equality of $P_P = L_L = A_A = V_V$ can be proven rather straightforwardly, but it is comforting to note that the ratios all have the same dimension, which is a simple ratio. Measurements that deal with the density of surfaces in a material must ultimately have dimensions of 1/length (area/volume), whether they are measured by counting points per length of line, or length of line per unit area, or something else. Likewise, densities of points and lines must have appropriate units. If your counting method is delivering numbers with an inappropriate dimension, you are almost surely making a serious error in the method being used for the determination.

Dimensionality is also important to keep from becoming confused about the effects of image magnification, shifts from inches to metric units, and so on. We have presented the relationships in this chapter (and will continue to do so) using line length, area and volume on the scale of the sample itself. The image magnification M is universally given as a ratio of lengths, so that 500X means that a line that is 500 mm (or any units) long on the image is 1 mm long on the specimen itself. The magnification must be used to adjust results from the measurement space to the specimen's dimensions, and it is important to keep in mind the units of the measurement.

For instance, V_V is dimensionless so magnification is not needed. P_L (the number of points per unit line length) must be multiplied by magnification, whereas L_3 (mean intercept length) must be divided by magnification. N_A (number per unit area) must be multiplied by magnification squared, and N_V (number per unit volume) must be multiplied by magnification cubed, to convert from measurements made with the image dimensions to those for the sample itself.

Chapter 4

Size Distributions

All of the parameters that were described in the preceding chapter are ones that apply to the sample as a whole. They tell us about global averages but not about the way that the structure is organized. An example of a very important parameter that is often desired to describe structures in which there are dispersed features in a matrix is the number of such features per unit volume, N_V. But this is not something that we can directly determine. Consider the cases shown schematically in Figure 1. The sections are identical, and show the same number and size of intersections, but the number and size of the 3-D features is quite different. Even if we restrict ourselves to the case of convex features, the number of intersections on our test section will depend on the size distribution and volume fraction of the features, as well as on their number. It is necessary to consider all of these things together.

Intercept length in spheres

Initially we shall consider spheres, whose diameter d completely defines (regardless of orientation) the probability that our test plane intersects a particular sphere; that probability is just d/w where w is the dimension of the cubic volume

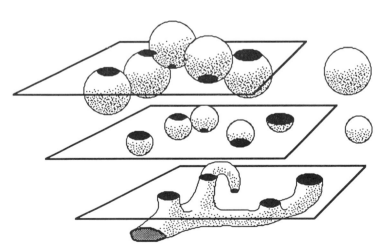

Figure 1. An example of different 3-D structures producing identical 2-D sections.

containing the sphere. This reflects the geometrical probability that a random plane section through the volume will hit the sphere. If there are n spheres randomly distributed in the volume, then the average number of profiles expected on a section (of area w^2) is just nd/w. If we divide by the area of the section, we obtain

$$N_A = N_V \cdot d$$

where N_A is the number of intersections per unit area on the section, and N_V is the number of spheres per unit volume in the specimen. When the spheres are not all of the same diameter, the average diameter must be substituted for d. (This may not be obvious, but it is true for any kind of sphere size distribution. It is a fundamental principle of stereology, and has also been rediscovered and republished several times, apparently first by Wicksell in 1925.) For other convex shapes, d is the "mean tangent diameter," the caliper dimension obtained by averaging over all orientations.

Of course, to actually determine the number of features per unit volume, even for the simple case of spheres of uniform diameter, this deceptively simple equation is not of much use. While it is certainly possible to count the number of intersections per unit area of test sections, to determine N_A, it is not so easy to get d (or a mean d in the case of varying sizes). There are some simplifying assumptions that can be made for uniform size particles (we might for instance suppose that after looking at a large number of section profiles, the largest one should be close to the diameter of the sphere), but the more general case is the one of usual concern so we will proceed to it now. It requires that we look into the subject of size distributions.

As is developed more fully in the appendix on geometric probability, it is possible, for any known shape, to calculate the expected distribution of lengths of intercept lines through particles of constant size. For the case of a sphere, as shown in Figure 2, the length of the intercept line depends only on the distance from the sphere's center, marked e in the drawing. The intercept length (for a sphere of unit radius) is just

$$L_i = 2 \sqrt{(1-e^2)}$$

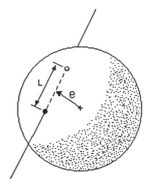

Figure 2. A line passing through a sphere. The intercept length L depends on the distance e from the center.

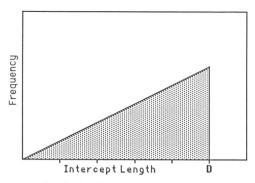

Figure 3. Frequency distribution for intercept lengths through a sphere of radius *D*.

Since all intercept lines that pass at a given distance lie on a cylinder, and the perimeter of the cylinder represents the fraction of all the intercept lines that pass through the sphere at that distance, it is possible to solve for the fraction of all intercept lines as a function of length. The result, which can be derived analytically without much difficulty, and can also be exhibited nicely by Monte-Carlo methods, is that the probability distribution curve for intercept lengths through a sphere is a straight line, as shown in Figure 3. The mean intercept length is hence *2/3* the diameter of the sphere.

Normally, instead of a continuous plot, we measure intercept lengths and count them in bins, each with equal width (range of length values). The resulting histogram plot for intersections, shown in Figure 4, is what we would measure as linear intercept lengths on random lines drawn on our planar test sections. The bin width δ is usually chosen to permit adequate counting statistics to define the shape of the histogram. In a structure with spheres of different sizes, the individual distributions would simply add, in proportion to the relative abundance of each size. Figure 5 shows the result, for several discrete sphere sizes.

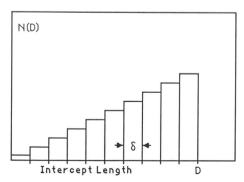

Figure 4. Histogram of the frequency distribution for intercepts through uniform spheres.

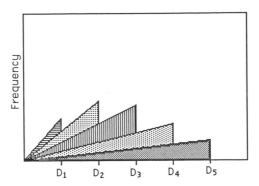

Figure 5. Superposition of intercept length curves for several sphere sizes.

A histogram with this kind of data can be taken apart graphically, as shown in Figure 6 (Lord & Willis 51): with the measured intercept length frequency plot, draw a series of straight lines from the origin (zero intercept length), through the center of the top of each bin in the histogram, to a construction line at the right of the plot. The lengths of the line segments marked out on the construction line represent the relative amount of the total frequency plot due to each size spheres, and so these values may be replotted as a histogram of sphere diameters as shown in Figure 7.

Note that the width of bins for the sphere diameter, in the plot of number versus sphere size, is the same as the width of the measurement bins used in the intercept length histogram, so depending on the need to count enough intercepts to adequately define the histogram (ordinary counting statistics), as many divisions as desired can be used to produce nearly continuous plots of sphere size distribution.

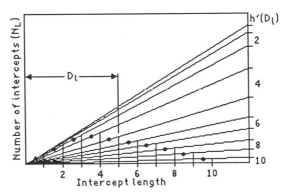

Figure 6. Extraction of the relative abundance of different sphere sizes from the measured histogram of intercept lengths.

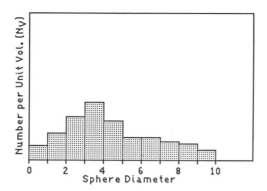

Figure 7. Plot of number of spheres versus diameter derived from the graph in Figure 6.

Non-Spherical Shapes

The sphere is not always a good, or even acceptable model for the shape of three-dimensional features. For instance, it cannot fill space, as grains in a metal do. But since the diameter of uniform spheres can be directly determined as 1.5 times the measured mean intercept length, and hence used to determine the number of features per unit volume (using the relationship at the beginning of this chapter), it is tempting to see how other shapes relate to the sphere.

Figure 8 shows a number of geometric shapes, including some regular polyhedra which can fill space (e.g. to model grains in a metal or cells in tissue), and some other shapes that are useful models for various types of real particles that occur in both biological and materials systems. For each, the table shows the ratio of the mean tangent diameter to that of a sphere with the same volume. It is noteworthy that

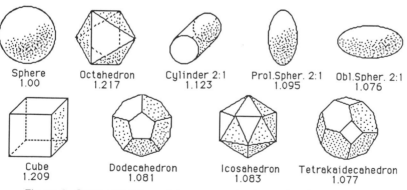

Figure 8. Some shapes useful as approximate models for grains and particles, with the ratio of their mean tangent diameter to a sphere of equal volume. (Hilliard 68)

many of the shapes, including polyhedra that may be reasonable models for grains in a material, have a ratio to the sphere of about 1.08.

The practical meaning of this is that since the ratio is quite insensitive to shape, we may expect real space filling grains and some other particles to also have a similar ratio. If we count profiles through the grains or particles on a section, we can calculate the mean diameter *d* from the intercept lengths using a spherical model, and multiply by 1.08 to obtain a size representative of a more realistic shape. Then the estimated number of three-dimensional features per unit volume is (Weibel 79)

$$N_V \cong N_A / (1.08 \, D_{spherical})$$

This may seem like an excessively crude approximation, but it is frequently as good as any other estimate which could be achieved, even at the expense of more effort. It breaks down when the grains or particles are grossly distorted or have shapes that change with size.

More exact methods, using tables derived by geometric probability, have been proposed to relate mean intercept lengths to the size and shape of a variety of three-dimensional generating shapes. If only the line lengths are recorded, it is necessary to choose a specific shape to determine a size distribution. The same shapes that were just discussed are typical ones that can be used. For convex shapes, various investigators (Weibel & Gomez 62, Dehoff 64) have proposed the relationships

$$N_V = [K / b] \, N_A^{3/2} / V_V^{1/2}$$

and

$$N_V = 2 \, g \, N_A \, P_L / P_P$$

where the variables N_V, N_A, V_V, P_L and P_P have their usual meanings, *K* is an adjustment for the range of size distribution, and the shape factors β and γ depend on the assumed shape of the 3-dimensional objects in the section (some representative values are shown in Figures 9 and 10, and Table 1). The parameter *K* takes into account the variation in size of the features from their mean. It is often assumed to have a value between 1 and 1.1 for many "real" structures, particularly ones that should all represent a single population (examples often cited are nuclei or other organelles in a single type of cell), but as shown in Figure 9, its exact value depends on a knowledge of the size distribution of the 3-D features. There are methods to obtain these, based on extensive point and intercept length counting.

Table 1. Shape coefficients for particle counting

Shape	γ	β
Sphere	6.0	1.38
Prolate Ellipsoid 2:1	7.048	1.58
Oblate Ellipsoid 2:1	7.094	1.55
Cube	9.0	1.84
Octahedron	8.613	1.86
Icosahedron	6.912	1.55
Dodekahedron	6.221	1.55
Tetrakaidekahedron	7.115	1.55

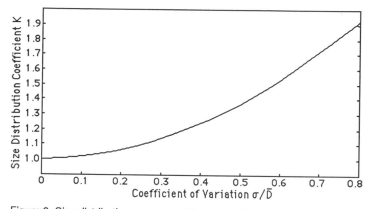

Figure 9. Size distribution parameter *K* as a function of the relative standard deviation of the mean diameter.

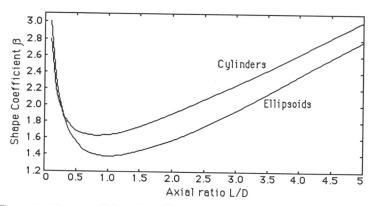

Figure 10. Shape coefficient β for ellipsoids and cylinders of varying axial ratio.

The β and γ shape factors vary only slightly for different shapes. Just as we saw before for the case of mean tangent diameter, since many varied shapes behave in a rather similar overall fashion, it may be acceptable to apply a modest correction to the results obtained using a simple sphere model.

Corrections for finite section thickness

We have so far restricted our discussion to measurements performed on images of planar sections through solids, as for instance would be observed by polishing metal samples and observing them with the incident light metallurgical microscope. For the biological thin section examined in transmitted light (or rock slices examined in the petrographic microscope, or ultra thin sections of anything examined in the transmission electron microscope), it is necessary to take into account the thickness of the section. As discussed in chapter 2, the effect of finite thickness will be to include more features from a minor second phase in the image than would be seen in a planar section, to hide their minimum dimensions, and sometimes to have overlap of several features present the appearance of fewer, larger particles.

For instance, the relationship we had before between number of features per unit volume N_V and the number of profiles per unit area N_A becomes

$$N_A = N_V (d_{mean} + t)$$

where t is the thickness of the section (Abercrombie 46). In the limit where this thickness is small and the area of the image is large with respect to the size of the features present, the results approach those for ideal planar sections.

Another interesting approach (Ebbeson & Tang 65) requires using two sections of different thickness, $t_1 > t_2$, and counting profiles on images of each. Then, without making any assumptions about particle size or shape, one can obtain

$$N_V = (N_{A1} - N_{A2}) / (t_1 - t_2)$$

but the difference between t_1 and t_2 must be fairly large, and the number of features counted also large, to obtain a result with useful precision.

It is also possible to utilize tables of coefficients like those to be introduced in the section relating profile size distributions to feature size and total N_V, to determine the number of features per unit volume in a finite section. The specific values depend on the section thickness, which must be known. Refer to Weibel (79) for detailed references. Cruz-Orive (83) shows a complete calculation method for this case.

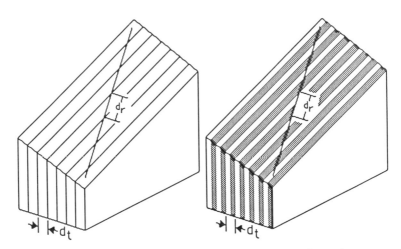

Figure 11. Diagram showing lamellar structures of one and two phases or structures with real spacing = d_t and the spacing along a random test line on an arbitrary surface = d_r.

Lamellae

Intercept methods are also appropriate for the measurement of lamellae. These are are particularly interesting in materials as a model for eutectic structures such as pearlite, twinned metal grains, banded rocks, or in biological systems for layers of distinguishable cells as may be found in plant tissue or muscle. Because these structures will not in general lie perpendicular to the section plane, the apparent spacing may be greater than the actual (perpendicular) spacing, as shown in Figure 11. Lamellar structures are also anisotropic, and may require analysis of their degree of orientation as will be discussed in Chapter 7.

The analysis of such structures from length intercept measurements on random lines (drawn at all orientations to the lamellae) depends on also determining the volume fraction of each structure or phase ($V_V = L_L$). The integration over all angles, as in the case of the needle problem discussed in the preceding chapter, gives $d_r = 2d_t$ where the subscript r denotes the mean spacing measured on the random lines, and t indicates the true normal spacing (of course, the mean random spacing is just $1/N_L$, the number of intercepts per unit length of test line).

When the thicknesses of lamellae vary, the volume fraction of lamellae with true spacing d_t is (Cahn & Fullman 56)

$$V_V(d_t) = 3\, d_r\, n_L(d_r) + d_r{}^2\, d\,(\,dn_L(d_r))\,/\,dd_r$$

where $n_L(d_r)$ is the number of measured intercepts of length d_r. It is usually necessary to have several hundred measured intercept length observations to produce a distribution function whose slope is well defined, to give numerical values for the derivative in the equation above. This relationship also holds true when the lamellae are not uniform (e.g. they may be tapered) or do not fill space. In fact, the method may even be extended to deal with isolated structures such as some membranes that have finite but varying thickness, in a matrix.

It is often important to determine the true lamellar spacing. As this is not likely to appear on a randomly oriented surface, what measurement can be made that will estimate it? If enough random intersections of the section plane with the microstructure are measured (this may require only one plane, if there are many islands of lamellae with different macroscopic orientation, as for instance in many eutectic structures in metals), then the surface area per unit volume S_V can be determined as was shown in Chapter 3. It is

$$S_V = 4\,/\,d_{r\,mean}$$

But for an ideal lamellar structure, the total surface area per unit volume would be

$$S_V = 2\,/\,d_t$$

(You may confirm this by imagining a cube of unit volume cut out with faces parallel to the lamellae.) Equating these two relationships, we would expect for an ideal lamellar structure to find a simple and fixed relationship between the two d values.

$$d_{r\,mean} = 2\,d_{t\,mean}$$

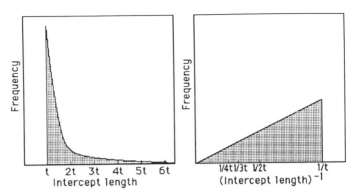

Figure 12. Intercept length distribution for a membrane of thickness t.

Note that this result is identical to that shown before for uniformly spaced lamellae. For real, as opposed to ideal lamellar structures, the coefficient may vary slightly from 2.0 (in pearlitic steels it has been measured as between 1.9 and 2.0), but this approximation will often be adequate for comparison estimates (Underwood 70).

A special case of lamellar structure is the isolated membrane or platelet of finite thickness, which we would usually like to measure. Again, we will assume that there are many such membranes, initially all of the same thickness, that can be sampled with a set of random lines drawn on a random section. For this case, the distribution of intercept lengths is as shown in Figure 12.

The plot shows that it is fairly likely that some value near the true thickness will be observed, but that there can be some very long intercepts (when the line is nearly parallel to the membrane). It would be difficult to use the methods described before for "taking apart" distributions of intercepts from a range of feature sizes to deal with this case. However, if the plot is instead made of number of intercepts versus $1/L$, a much simpler distribution shape is obtained (Gunderson 78), which can be converted to a distribution of number of features versus membrane thickness just as the sphere intercepts were.

Measurement of profile size

It is also possible to determine a distribution of sphere sizes in a matrix by measuring the size of the circular profiles cut through the spheres by the planar

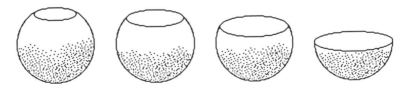

Figure 13. Different circle sizes produced by sectioning a sphere.

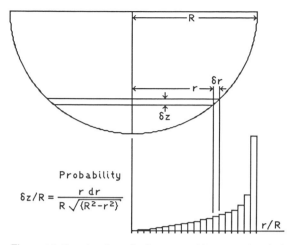

Figure 14. Construction of a frequency histogram for circle sizes obtained by random sectioning of a sphere.

sampling surface. Figure 13 reminds us that every section through a sphere is a circle, with a maximum size having the same diameter as the sphere. From the approach of geometric probability, consider the different profile sizes that can be obtained on the sphere, depending on where the section happens to hit it. A histogram of the frequency of observing particular size circles, for random sections, can be derived either by analytical or Monte-Carlo means. Using a simple analytical approach, we can see (as shown in Figure 14) that the probability of obtaining a particular circle diameter is related to the vertical thickness of a slice of the sphere with that size. This allows us to construct the frequency histogram shown.

It is not hard to show, either analytically or graphically, that the mean diameter of the circles is related to the diameter of the generating sphere by

$$D_{sphere} = 4 / \pi \, d_{mean}$$

and the shape of the frequency distribution curve in terms of the circle diameter d and the sphere diameter D, is given by

$$P(d_{circle}) = d / \{D \, (D^2 - d^2)^{1/2}\}$$

Now, if there are several different sizes of spheres all present in the sample, a particular size circle could result from intersection of the planar section with many different sphere sizes. The frequency distributions for each size will add, giving rise to a complex size histogram as shown in Figure 15. This is what we would actually measure on the sections. The problem we are usually faced with is to deconvolute this complex distribution of circle sizes into the distribution of sphere sizes, from which (among other things) the mean diameter and hence the number of features per unit volume can be obtained.

This could be done graphically, by subtraction, as we did for the linear intercepts, since the largest circle diameters could only come from the largest spheres. Then the distribution of circle sizes from spheres of that size, scaled to match the number of the largest circles observed, could be subtracted from the distribution. The remaining histogram would then be processed in the same way for the next smaller circle size, and so on until all of the sphere sizes are accounted for, and the number of each size obtained. Unfortunately, this method is not usually practical, because the individual distributions are quite nonlinear, with only a few expected small profiles, and the limited statistical precision of the number of counts in the individual size classes in the histogram causes the errors to propagate so that quite wrong values (even negative numbers of some sphere sizes) may be obtained.

A better approach is to solve a set of simultaneous equations to relate the sizes of spheres to the measured histograms. This has been done for the case of spheres by several researchers, first by Saltykov (67). The solution leads to a matrix of coefficients (see Table 2 for the coefficients of Cruz-Orive (76), which are recommended for practical use) which can be used to multiply the numbers of circles in the measured histogram to obtain the corresponding distribution of spheres.

This matrix is obtained by first calculating, from geometric probability, a frequency distribution table for N_A values in each size class due to three-dimensional objects of each size, N_V. The largest intersections come only from the largest objects, but the smallest intersections can arise from any size object. Summing the frequency values for all contributing object sizes produces a set of equations for N_A in each size class

$$N_{Ai} = \alpha'_{ij} N_{Vj}$$

This matrix can then be inverted to obtain the α-matrix, which solves for N_V as a function of N_A. Note that if this is done for spherical objects, the columns in the

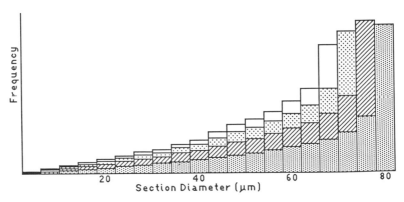

Figure 15. Size histogram of circles measured from intersections on several sphere sizes is the sum of the individual distributions shown by the differently shaded bars.

Table 2. Matrix of coefficients used to calculate the size distribution of spheres.

Profile Size Class	1	2	3	4	5	6	7	8
				Particle Size Class				
1	+2.000							
2	−0.68328	+0.89443						
3	+0.08217	−0.47183	+0.66667					
4	−0.02799	+0.04087	−0.39550	+0.55470				
5	−0.00048	−0.02528	+0.02903	−0.34778	+0.48507			
6	−0.00298	−0.00393	−0.02416	+0.02308	−0.31409	+0.43644		
7	−0.00126	−0.00427	−0.00516	−0.02293	+0.01947	−0.28863	+0.40000	
8	−0.00094	−0.00234	−0.00488	−0.00565	−0.02176	+0.01704	−0.26850	+0.37139
9	−0.00062	−0.00167	−0.00288	−0.00512	−0.00582	−0.02071	+0.01527	−0.25208
10	−0.00045	−0.00117	−0.00208	−0.00315	−0.00518	−0.00584	−0.01977	+0.01393
11	−0.00034	−0.00087	−0.00150	−0.00230	−0.00327	−0.00509	−0.00571	−0.01893
12	−0.00026	−0.00066	−0.00113	−0.00169	−0.00242	−0.00332	−0.00509	−0.00571
13	−0.00020	−0.00051	−0.00087	−0.00129	−0.00180	−0.00248	−0.00332	−0.00501
14	−0.00016	−0.00040	−0.00069	−0.00101	−0.00139	−0.00186	−0.00250	−0.00330
15	−0.00013	−0.00033	−0.00055	−0.00080	−0.00109	−0.00145	−0.00189	−0.00250

Profile Size Class	9	10	11	12	13	14	15	Total
				Particle Size Class				
1								+2.000
2								+0.2115
3								+0.27700
4								+0.17208
5								+0.14056
6								+0.11435
7								+0.09722
8								+0.08436
9	+0.34816							+0.07453
10	−0.23834	+0.32880						+0.06674
11	+0.01287	−0.22663	+0.31235					+0.06042
12	−0.01818	+0.01200	−0.21649	+0.29814				+0.05519
13	−0.00560	−0.01751	+0.01128	−0.20761	+0.28571			+0.05080
14	−0.00491	−0.00549	−0.01691	+0.01067	−0.19973	+0.27472		+0.04705
15	−0.00327	−0.00481	−0.00538	−0.01636	+0.01015	−0.19269	+2.6491	+0.04382

original matrix are just as shown in Figure 14 above, but that it is equally possible to assume some other generating shape, for instance a cube or ellipsoid of revolution. For any shape other than a sphere, it is necessary to rotate the object through all possible orientations to obtain the probability curve of intercept areas. With the α-matrix coefficients, the procedure to follow comprises the steps of measurement, plotting and calculating detailed below.

1. Measure: Record a size distribution of profiles in about 10-15 linear size categories. A count is recorded in a size class if the measured profile fits between two circles which are the limits between the class and its neighbors. The mean diameter of the two circles is taken as the characteristic size for the class. The reference area is determined as the area of the image where features are measured. A procedure to

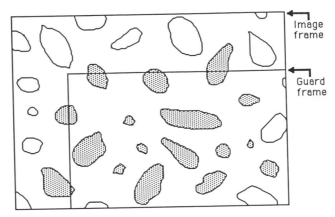

Figure 16. Application of a guard frame to an image. Only the shaded
features are measured or counted, and the measured area is that within the
smaller frame.

guard against edge effects must be invoked so as not to bias the data against large
profiles which are more likely to intersect the edge of the image. Two methods are
available.

A: Draw an internal frame within the overall image, as shown in Figure 16, and
measure in their entirety those features which lie wholly within the frame, or
which lie partially within the frame and do not intersect the two forbidden outer
edges. The reference area is the area of the inner frame. This method reduces the
useful area of the image that can be measured (in principle, by 75% since the
guard region must be as large in each direction as the measurement region), but is
probably the most widely employed to eliminate bias against large features (small
ones are less likely to intersect an edge, and so would be counted in greater
proportion if no such guard frame was used, but edge-touching features which
could not be measured were bypassed).

B: Use the entire image area as the reference area. Measure all features which lie
within the image (do not touch any edge) but for each, count in the histogram a
number greater than 1 given by

$$(W_x \cdot W_y) \, / \, \{(W_x{-}F_x) \cdot (W_y{-}F_y)\}$$

where W_x and W_y are the widths of the image in the X and Y directions, and F_x
and F_y are the widths of the features (the caliper or Feret's diameter) in the X and
Y directions. This is equivalent to constructing a different measurement frame for
each feature. Small features, which are unlikely to touch an edge, are counted
about 1; but large features are counted much more than one to compensate for the
fact that more of them are lost due to touching an edge. This method is equivalent
to the first method, and is especially preferred with machine counting methods
because it is more efficient, the various widths are usually determined anyway,

and the arithmetic is easily handled (bin contents become real numbers rather than integers).

2. Plot: Construct the histogram of number of objects as a function of size (the diameter of a circle with the same area as the observe profile), and then fill in the missing portion of the histogram (as recommended by Weibel 79). This is necessary because the smaller profiles will be lost or undercounted due to finite resolution of the images, loss during sample preparation, and so forth. It is permissible because the smaller size features in any histogram are rarely important in defining the structure of the material. For instance, even a very large number of small features contribute less total volume fraction that a few large ones. The small features can be important in surface/volume calculations, but not in this instance. Weibel suggests, in fact, that it is often best not to even attempt to measure or count the smallest quarter or third of the histogram, but simply to complete it as shown in Figure 17.

In this example, a visual estimate of the smoothed shape of the histogram is determined. Then the maximum is located and a point is marked at half the visually smoothed height of the distribution at this size. From this point, a straight line is constructed to the zero size class. Where this line predicts a number of counts greater than that actually observed, you are expected to assume that the difference is due to undercounting features that were not seen, and use the higher value. The histogram should contain a few hundred measured feature sizes to yield a reasonable statistical result for the final size distribution. This is equivalent to using the measured histogram for the N_A values, finding that multiplying by the α-matrix gives negative values for some N_V's (for the smallest sizes), and ignoring the negative values as not corresponding to physical reality.

Note: this is not to be taken as a general approach to completing histograms, especially when dealing with projected images, or types of data in which the smaller sizes contribute something of importance to the parameters of interest. In that case, if it is suspected that small features are being lost due to finite image resolution, it is necessary to obtain more images at higher magnification, make measurements on

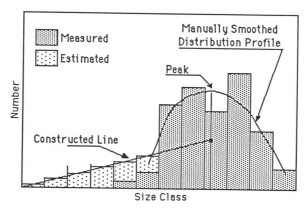

Figure 17. Schematic showing the method of completing a histogram for missing small features.

those images, and combine the two histograms (one for small features and an adjoining one for larger features that are too big for practical measurement in the high magnification images). The two histograms must be scaled in proportion to the reference area of the images measured at each magnification, which will also in general give rise to fractional numbers of counts in some size classes.

3. Calculate: Use the table of coefficients to multiply the numbers of counts in each size class and add, to obtain a histogram of sizes of spheres in the volume. The numbers in the column marked "total" can be used to multiply by the number of counts in each profile size class and sum the total number of particles per unit volume. The overall histogram can be examined visually or by calculating backwards the number of small circles which the resulting histogram should generate, to check the reasonableness of the assumed corrections to the circle histogram at small sizes, if this correction was made. The sum of particles in each size class j is accumulated from the number of profiles in each profile size class i using the α_{ij} coefficients from the table (blanks are zeroes):

$$N_{Vj} = S\, a_{ij} N_{Ai}$$

$$N_{Vtotal} = S\, a_{i\, total} N_{Ai}$$

As an example of this method, Figure 18 shows an image of bread, in which the spherical features are the pores, and the sample is obtained by cutting stale bread and darkening the cut surface with a felt ink pad. The techniques of Chapter 3 provide the information that the volume fraction of the pores is 80.3%, and the mean thickness of the dough is 0.022 mm. Values for these parameters are important in the quality control of breads. Measurement of the pores produces a histogram as shown in Figure 19, using the diameter of the circle for the size class. If these are converted from N_A to N_V as just described, a rather different histogram is obtained, as shown

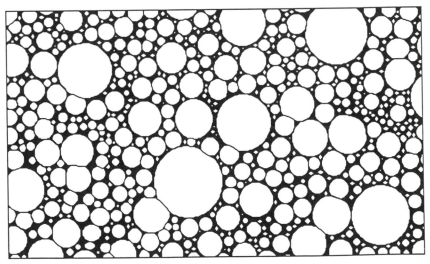

Figure 18. Pores in bread (area of image is about 10 square millimeters)

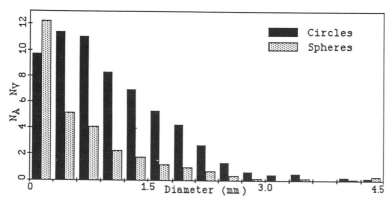

Figure 19. Frequency histograms for N_A (number of circles per unit area, based on the circle diameter) and N_V (number of spheres per unit volume, based on the sphere diameter) for the pores shown in Figure 18.

superimposed on the same plot. (There are over 300 pores in this one image, and more are needed to satisfactorily specify the details of the histogram for larger pores, so rather than perform these measurements by hand, they were actually done using the automatic methods described in the following chapters.)

Non-spherical particles

Alas, many real structures are not adequately modelled by distributions of spheres in a solid matrix (although a surprisingly large number of problems can be handled with this rather extreme simplification). By using geometric probability, and integrating (or using Monte-Carlo methods to sample intercept profiles), it is possible to obtain matrices of coefficients that will convert any histogram of two-dimensional sizes to a histogram of sizes of the three-dimensional generating shapes. Specific cases that have occasionally been used include cylinders (short, squat cylinders are good models for plate-like second phase or precipitate particles, and long slender ones can be used to fit needles or other acicular shapes), and other polyhedra. The problem with these shapes is the difficulty in describing the size and shape of the two-dimensional profiles that appear in the section image.

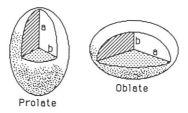

Figure 20. Prolate and oblate ellipsoids.

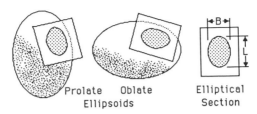

Figure 21. Sections through prolate or oblate ellipsoids are ellipses.

A much more general shape is that of an ellipsoid of revolution. Figure 20 shows ellipsoids that are prolate (generated by rotating an ellipse around its major axis) and oblate (generated by rotation around the minor axis). These shapes can also be used to model a wide variety of roughly convex shapes for second phase particles.

Furthermore all of the plane sections through the ellipsoids produce elliptical profiles (Figure 21). These may be characterized by two dimensions, such as the length of the major and minor axes (or any other two dimensions, such as the area and aspect ratio, for that matter). A standard method for using these shapes to describe features in a matrix, and determine their size distribution, was developed by Dehoff (62). This restricts the 3-D shapes to ellipsoids of a constant shape (axial ratio), but with varying size (diameter).

Then the expression used to convert the measured size histogram of 2-D profiles to 3-D features is a modification of the method described before for spheres:

$$N_{Vj} = \{ k(q) / \delta \} \; \Sigma \; \alpha(i,j) \, N_{Ai}$$

where j indicates the size class of the 3-D objects, i is the class number of the profile size, the α factors are from the table above, δ is the size increment, and the $k(q)$ are specific shape factors for the ellipsoids of axial ratio q. These have been calculated for various shape ellipsoids, as shown in Figure 22.

Note that it is necessary to decide whether the generating shapes are oblate or prolate, either of which can produce the same elliptical profiles on the plane section image. In most real instances, it is possible to make this decision based on independent information about the specimen. If it is not known from other sources which shape is appropriate, then if there are enough profiles in the section to give a statistical representation of all sizes and shapes, the following logic may be applied (see Figure 23): If the generating particles are prolate (football shaped), then the diameter of the most nearly equiaxed section(s) will be similar in size to the width of those with the highest aspect ratio. Conversely, if the generating particles are oblate (discus shaped), then the diameter of the most nearly equiaxed section(s) will be similar in size to the length of the most elongated profiles. This comparison should be made with only the 5-10% of the features with the lowest and highest eccentricity. The parameter q is taken as the ratio of minor to major axis from the most elongated profile(s), and used to obtain k from the graph. The method is then just like the conversion for spheres.

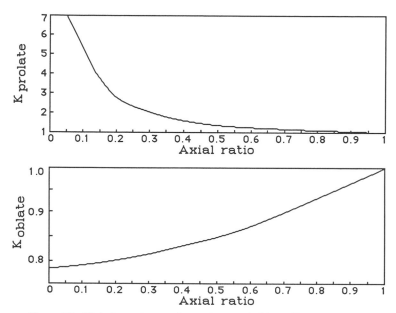

Figure 22. *K(q)* shape factors for prolate and oblate ellipsoid, versus the axial ratio of the generating ellipse.

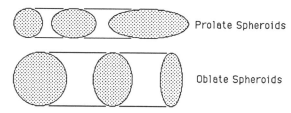

Figure 23. Comparison of elliptical profiles to determine if generating ellipsoids are oblate or prolate.

It is of course implicit in this method that all of the three-dimensional objects are assumed to be of the same shape. In cases where there are mixtures of features in the 2-D image, due to the presence of different shape generating objects, it is often practical to separate them into distinguishable populations of features based on some aspect of their appearance, such as color or density in a light image (perhaps with an appropriate etching or staining procedure), or using backscattered electrons or X-rays in the SEM. Then each family of features can be analyzed separately.

In principle it should be possible to extend this method of analysis to convex features which do not all have the same shape. After all, in the two-dimensional section, two independent parameters (such as length and breadth, or area and eccentricity) of the intersection profiles can be obtained. From these data, calculation of both the size and shape distribution of the generating ellipsoids should be

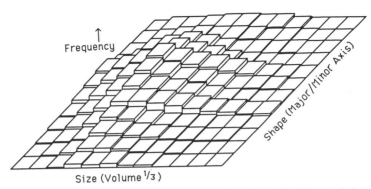

Figure 24. Histogram of particle size and shape based on ellipsoid of revolution
model.

possible. Likewise, it should be possible to measure intercept lengths in two
perpendicular directions and from these data determine the size and shape distribution
of the generating shapes. If a shape for the 3-D objects to be used as a model is
defined, then geometric probability can be used to calculate the necessary matrices of
coefficients to relate the number of measured intercept lengths or profile dimensions
in each size class to the number of generating shapes in each size and shape class.
Then this 4-dimensional matrix of coefficients can be inverted to produce the α-
matrix used to compute the number of objects per unit volume in each size and shape
class, as a function of the number of outlines per unit area in each size and shape
class.

$$N_{Vij} = \Sigma\, \alpha_{ijkl}\, N_{Akl}$$

Figure 24 shows the result, in the form of a 2-dimensional histogram of size
and shape for the generating particles. It is important to select a useful shape model
that is easily sampled for intercepts, such as ellipsoids of revolution with varying
eccentricities. The major difficulty is a statistical one. The population of sparse bins
in the histogram, especially for the more extreme intercept lengths or section profile
sizes, which are geometrically of low probability but important to the specification of
feature shape, propagate a substantial error into the final answer.

Chapter 5

Computer Methods

Up to this point we have dealt with manual techniques of image measurement, and have presented methods by which we can obtain parameters that describe the three-dimensional structure with a minimum of measurements, and easily obtained ones, on two-dimensional images (usually of sections). The advent of computer-based measurement systems has not simply speeded up the calculation of these same parameters, but has changed somewhat the types of things that are measured and the ways in which some familiar quantities are determined. As a simple example, whereas with manual methods the volume fraction would be determined from P_P with at most a few hundred points, a single video image will typically have over 50,000 points, and the fraction of these that lie in a particular phase is more nearly equivalent to A_A as it would be determined in the manual method by cutting and weighing (because large numbers of points lie within each feature). Also, the length of boundaries on surfaces is usually determined not by counting point intercepts with random lines, but by connecting points where each scan line has intercepted the boundary, or where a human has outlined regions by drawing. The lengths of these polygons are taken as direct measures of the length of the boundary, and from them the surface area per unit volume can be determined as was described before.

Semi-automatic (human delineation)

There are two principal approaches to obtaining the necessary raw image information in the computer. When contrast in the image isolates the features or objects of interest from each other and their surroundings, automatic discrimination can be used. More complex images (in the sense that brightness values vary on both the features of interest and the rest of the image, over the same or overlapping ranges) generally require either extensive processing (discussed later in this chapter), or human outlining of the features. Figure 1 shows an example: an SEM image in which the eggs which are to be measured were collected (along with considerable debris) on a filter paper. The range of brightness on the curved eggs is no different than that on the remainder of the image. Extensive "feature recognition" software might be able to isolate the eggs from the remainder of the image (after all, vehicles can be automatically recognized and located, and in some cases even identified as to type in satellite photos), but this requires considerable computation in substantial computers. It is more efficient for our purposes to use the computer as a measurement aid and data collector, but to mark or delineate the features by hand.

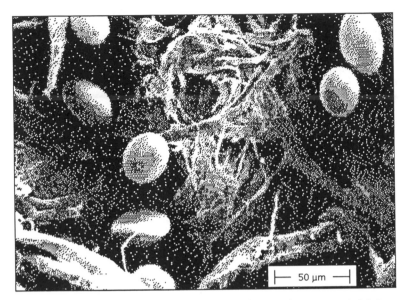

Figure 1. Scanning electron micrograph of nematode eggs and debris embedded in filter paper.

The semiautomatic method allows the human operator to mark boundaries, features, and so forth on the image using some device whose location the computer can directly record. A very common example is that of so-called graphics tablets or digitizers, on which photographs (or maps, etc.) are placed. A stylus can be moved over the image, and the computer determines its location by making ranging measurements, often by sending high frequency electromagnetic radiation through a grid in the tablet and measuring the time delay until the signal is picked up by the stylus. There are acoustic, resistive, capacitive, mechanical and other schemes to perform the ranging, and while most devices utilize cartesian (X,Y) coordinates directly, some use the distance from two separate points and calculate the X,Y position. All of this is usually transparent to the measuring program, and certainly to the operator.

Most tablet digitizers have the ability to determine the stylus location to better than 0.1 mm, out of a distance range of 25 or 30 cm (or more – there are digitizing tablets used in map reading that have active regions over 6 feet wide). The reading precision is generally better than the human can reproducibly position the stylus. Styli may be shaped like writing pens (convenient for tracing lines) or like crosshair reticles with magnifying glasses (excellent for marking specific points), and usually have a pushbutton to mark locations, or can be pressed against the tablet surface to signal to the computer.

For direct measurement without using photographs, another approach that is growing in popularity is to attach a video camera directly to a microscope, or for macroscopic objects to use a well lighted camera stand. The conventional TV image then appears on a viewing monitor, along with a cursor that the operator can manipulate through the computer. This may be done with an *X-Y* input device like the tablet digitizer, or using alternate devices such as a mouse or light pen. The latter consists of a light-sensitive diode housed in what looks like a ball point pen, which can be pointed at or held against the video monitor. When the continuously scanning beam in the cathode ray tube passes by the light detector, an electronic signal is generated which is fed back to the computer, which uses the time since the last beam retrace to determine the *X* position of the pen, and the time since the frame was started to determine the *Y* position. While these devices are intuitively obvious to use, they suffer from several practical problems: the pen (and the hand holding it) get in the way of the user's view of the image, and the position sensitivity falls off badly when the pen is pointed at dark regions of the image.

A mouse is a small box with a ball in its underside which rolls as the box is moved in *X* and *Y* directions on a desktop or other surface. Encoders follow the ball's motion and send signals to the computer, which can then determine how the user has moved it and move the cursor on the video screen accordingly. Recent developments in computer technology seem to indicate that the mouse is an extremely natural device to be used for pointing at objects, and they are becoming common. Their chief drawbacks are that they report only relative motion, and not absolute position (the mouse can be picked up and moved to a new location without the computer knowing it), and while they are very good for pointing at things, it is less easy to use them to draw smooth lines around shapes on the video image, while looking at the image and moving the mouse on the desk (although this eye-hand coordination is quickly learned with some practice).

Regardless of what the input device is, and whether the image is a photograph which the user is marking on, with a stylus and digitizing tablet, or is a live video image being outlined directly on the screen, the computer receives a series of coordinates as the user marks outlines around features of interest. The resolution of point position on the video image is limited to the number of lines on the video display (about 200 lines vertically and 250 points horizontally with conventional US television, which is interlaced), because those are the only positions where the cursor can be placed. However, this is the same resolution as that with which the image itself is shown on the screen, so if features of interest can be viewed in this way, they can be satisfactorily measured. The chief effect of the more limited image resolution is the need to work at more different magnifications than with photographs.

Even with continuous motion on the part of the user, the coordinates are separate and are determined quickly enough that most of the time the programs are waiting for the user to move to another point. These points form a series of line segments, which may each be so short that they give the appearance of continuous smooth lines. The line length can be summed up using the Pythagorean theorem:

$$L = \Sigma \{(X_i-X_{i-1})^2+(Y_i-Y_{i-1})^2\}^{1/2}$$

We have already seen how to use this boundary length to determine surface area per unit volume, contiguity and so forth.

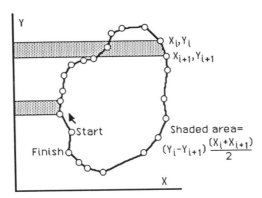

Figure 2. An outline polygon and the trapezoids summed to determine the enclosed area.

Features may also be completely outlined to determine their area. The first thing that must be done in this case is to deal with the fact that the user's line may not exactly close on itself. Usually, the programs will recognize a small error in closure as accidental and complete the polygon with a straight line segment from beginning to end. Then the area within the outline is determined as shown in Figure 2, as the algebraic sum of trapezoids from some arbitrary reference line (typically whatever the input device treats as its zero point). The summation proceeds all the way around the polygon from the first point to the last (which coincides with the first one).

$$A = \Sigma (X_i + X_{i-1}) \cdot (Y_i - Y_{i-1}) / 2$$

Since on one side of the polygon the difference in successive Y values is positive, and on the other side it is negative, the algebraic sum gives only the net area within the outlined region, regardless of where the reference line was taken. By taking the absolute value of the area, it is unimportant in which direction the outline was drawn, but an outline drawn like a figure "8" that crosses itself will have a net area that is small (the difference between the two loops, which have opposite signs). It is also possible to use this reversal of direction to obtain net areas for figures that have central holes, as shown in Figure 3.

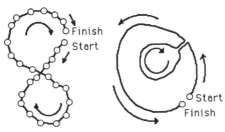

Figure 3. Drawn paths that give the net area difference between the two loops.

Feature-specific parameters

Once a feature outline (the series of typically a few hundred points making up the polygon) has been entered into the computer's memory, there are a number of parameters that can be obtained. For instance, in addition to the area and perimeter, described above, we often want to know the Feret's diameters (the caliper or projected diameters, Feret 31). For the Feret's diameters in the X and Y directions, this simply requires sorting through the coordinates for the largest and smallest values of X and Y. The difference, as shown in Figure 4, is then the desired value. These may be used to correct for the effects of image edges, which (as was described in Chapter 4) reduce the number of large features that can be measured because large features are more likely to intersect edges than small ones are.

Often more useful than the X- and Y-Feret's diameters are the longest and shortest values, which we shall call the length and breadth. Using a pre-calculated table of sine and cosine values for a series of angles, perhaps every 10 degrees, the coordinates in the outline can be quickly converted to other, rotated coordinate systems in which the largest and smallest values of X' and Y' can be found. Then the greatest difference in values, which is the largest caliper diameter or length, and the smallest difference or breadth can be determined, as shown in Figure 4.

The error due to the finite set of angles that are used in the transformation can be readily estimated. If the feature's true longest dimension does not correspond to one of the angles, the length will be underestimated as the cosine of the angle difference, and the largest the angle difference can be is half of the step used in the search. So for a 10 degree step, the worst case error would be cos (5 deg) or 0.996 (the lengths determined would be 0.4% low, probably a negligible error in most cases considering the finite accuracy of marking the points). Even for a 22.5 degree step (requiring only 7 coordinate transformations), the worst case error in length is only 1%.

For the breadth, the problem can be more serious. As shown in Figure 5, if the profile is very long and skinny, the perceived breadth can be greater than the actual minimum caliper diameter by any amount. Furthermore, this parameter is apt to be considerably larger than the actual width of the feature at any one location (often described, at least for the case of projected outlines of particles, as the size hole

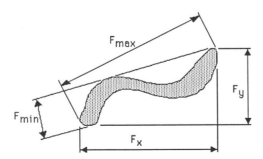

Figure 4. The Feret's diameters of a representative feature outline.

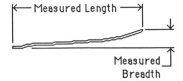

Figure 5. The error in breadth measurement depends on the length of narrow features and the angle difference from the measuring direction.

through which the particle could pass, if it were turned correctly on its way through (see Figure 6). This last parameter is extremely difficult to determine, requiring an exhaustive search through the distances from each point on the perimeter to every other, and consequently it is rarely used).

We will see later on that the length and breadth, defined as the longest and shortest caliper or Feret's diameters determined over some series of angles, will be useful in fitting ellipses and ellipsoids of revolution to the profiles. The underlying principles are the same as for the manual methods described previously, but in this case we have the actual values for the length and breadth, which require more effort to determine by manual measurement techniques.

It is also straightforward to record the angle of orientation of the length direction (or at least the nearest angle at which the coordinate transformation was made). Plots of the number of features, or their total length, as a function of orientation angle are often very useful in studying preferred orientation in materials, alignment of three-dimensional objects, and in for diverse images as porosity in compacted soils (the pores tend to align themselves at 45 degrees to the compression direction), root hairs on plants, geological fault lines from maps or aerial photos, and so forth.

A word of caution in dealing with these angle values: most of the parameters that we measure, either using manual or automatic techniques, can be treated

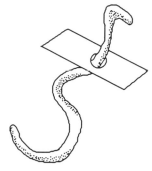

Figure 6. The sieve hole size through which a feature will pass with enough agitation to find an optimum orientation may be much smaller than any readily measured dimension.

straightforwardly to determine the mean and other statistical parameters. This is not so with angles. The average angle from the values 80, 85 and 110 is 91.7 degrees, determined by summing the three values and dividing by three. But the average angle from the values 5, 10 and 170 degrees is not 61.7 deg, but rather 1.7 degrees. It is important to recognize that either end of a line is equivalent, and that 180 degrees and 0 degrees are the same.

Another note: When we use the length and breadth to describe an ellipse to be fitted to the feature outline, we generally picture the breadth as being perpendicular to the length (it certainly is for the ellipse). But there is nothing in our method that ensures this to be so in general. For example, a square with side 2 would have a length of 2.828 (the diagonal) and a breadth of 2, and the two would be at an angle of 45 degrees to each other. Furthermore, if the square was perfect, there would be two equivalent directions for the length, and either might be chosen depending on the order in which the directions were sorted. If there were many such square figures, all aligned with edges horizontal and vertical, but with slight errors in the coordinates (as would be expected from the measurement techniques), about half of them would show an orientation angle of 45 degrees and half 135 degrees. Only by examining the images would we recognize the redundancy of this information.

The location of a feature is usually taken as the center of gravity of the profile, which is the point at which it would balance if cut out of rigid, uniform density cardboard. Mathematically, this point is determined from summation of the second moment of the area, or

$$CG_x = \Sigma \{(X_i + X_{i-1})^2 \cdot (Y_i - Y_{i-1})\} \ / \ area$$

and similarly for the y-direction. For convex shapes this point lies within the outline, but this is not necessarily so for concave profiles.

Automatic image discrimination

The principle reason that human marking of the boundaries between phases, or the outlines around features of interest, is required is that many types of images are complex and fully automatic methods of feature discrimination are not able to reliably distinguish all, and only, the right features. We carry very complex image processing and recognition computers between our ears!

Generally, the only tool that automatic instruments have available to distinguish features from their surroundings (or from each other) is brightness. If the use of special illumination (eg. colored or polarized light) or special images (eg. backscattered or X-ray images on the SEM) can produce grey-scale contrast that is different for the phase or features to be measured, then automatic image measurement is much faster than manual or semiautomatic techniques.

This discrimination is achieved by having the computer read the brightness of every point in a raster scanned image, either using a conventional TV scan or a similar device. For instance, some special cameras have more lines than a conventional video camera, while a scanning electron microscope scans much more slowly, and some scanning densitometry tables physically translate a film over a light source while a photomultiplier tube measures the light intensity. In any case, the computer determines the position of the spot either by timing or measuring voltages, and the brightness by measuring a voltage. The voltage is determined using an "ADC" (analog to digital converter), which produces a number that is proportional to the voltage. These devices are characterized by their range, expressed as the number of bits; most are 6 or 8 bits, meaning they can determine the voltage and hence the brightness to one part in 2^6 (64) or 2^8 (256). Since the human eye is not capable of distinguishing even 64 levels of brightness, the only real use for the larger range ADC's is to permit absolute levels of image brightness to vary without having to make any offset adjustment so the measured range lies within the capability of the digitizer.

When the image is read by the automatic system, it amounts to a series of discrete points (or "pixels," from "picture element"), arranged in a raster pattern like squares on a checkerboard (and usually the pixels are square, with the same spacing vertically as horizontally, as this greatly simplifies subsequent data analysis), with each point represented by a number describing its brightness (for instance, 0 to 63). The process is generally rather fast, ranging from one video frame (1/30th of a second) to a few seconds, except in cases like the SEM where the instrument itself produces data on a much slower scale.

In some cases the entire grey-scale image is recorded in memory so that the computer can access it (image processing, as contrasted to image measurement, will be discussed later in this chapter). But this requires rather large amounts of memory (a 256x256 point image, with one byte per point, would occupy 64K of storage space, and the size grows as the square of the number of points in each direction). Consequently, the image is often discriminated as it is acquired, so that only one bit is required for each pixel.

Discrimination simply means that the brightness of the pixel is compared to upper and lower level cutoff values that have been set (generally by the human operator) to separate the pixels that lie in the features of interest from their surroundings. When an entire image can be scanned beforehand, a histogram of brightness can be used to help set the discriminator levels. Sometimes simple trial and error, turning knobs that define the discriminator settings while viewing the "binary" image formed of white pixels where the brightness is within the settings and black where it is not, is fast and acceptable. In some cases, automatic algorithms have been devised to choose reproducible cutoff points.

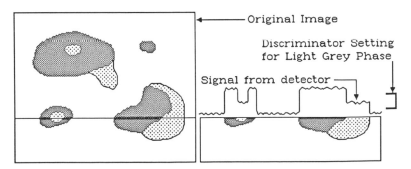

Figure 7. A grey scale image, and the signal produced as it is scanned.

Figure 8. The binary image produced by the discriminator setting shown in Figure 7.

Figure 7 shows a representative image, with the brightness signal along one scan line. The discriminator setting shown will produce a binary image as in Figure 8. If the entire image is scanned, and the individual pixel brightnesses counted in a histogram, the result might look like Figure 9. This would be useful in choosing the best discriminator settings to select one phase of interest.

The discriminator settings are important because they control the size of the features. In the example of Figures 7-9, there are three obvious peaks in the brightness histogram which correspond to the two phases, and to the matrix. They are well separated. However, some pixels lie on the boundary, and their brightness will be intermediate, depending on just how far they lie on one side or the other of the interface. Setting the discriminator at a value closer to the center of the peak will reject these points, and will make all of the features somewhat smaller than a wider setting will. For consistency (precision, as opposed to accuracy, as was discussed in Chapter 1), it is often recommended that the cutoff be placed halfway between the two average brightness levels on the isolated phases or objects.

The grey scale image is an array of numbers, as shown schematically in Figure 10 (this is only a tiny fragment of a typical image, with 400 pixels in a 20 x

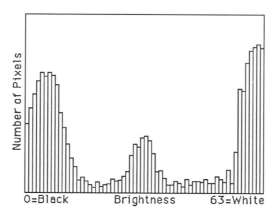

Figure 9. The brightness histogram corresponding to the image in Figures 7 and 8.

20 array, and each pixel brightness has only 10 possible values instead of the more typical 64 or more).

A binary image obtained with discriminator settings that included all pixels with brightnesses between 6 and 9 (inclusive) would look like Figure 11. This requires only one bit (0 or 1) per pixel, instead of one byte of storage, so the image takes up less space in the computer.

Even the binary image takes up more space than is really essential. In order to make measurements on the features, what is really needed is the perimeter points, just as were obtained before from the semi-automatic entries. This can be obtained by checking each scan line on the image (either vertically or horizontally, whichever is

```
2 2 3 2 2 3 3 4 4 4 2 4 3 3 4 3 4 4 2 5
1 2 2 1 3 2 3 3 3 4 4 3 2 3 3 5 5 3 4 3
1 2 3 1 2 8 3 4 3 2 2 3 4 3 4 3 3 3 5 8
2 1 2 8 7 5 7 7 2 4 3 3 3 3 5 4 4 3 5 9
3 1 9 6 5 5 5 6 9 2 4 3 4 3 3 3 4 5 9 9
1 3 7 6 4 2 3 5 6 4 3 5 4 4 3 5 5 4 8 7
3 9 7 5 1 0 1 3 5 9 2 3 5 3 5 5 5 7 9 9
3 2 7 4 4 2 2 4 8 3 4 3 3 3 4 5 4 9 9 7
1 3 9 7 6 3 5 5 8 3 4 4 4 5 3 4 8 9 8 7
3 3 2 8 7 7 6 8 3 4 4 5 5 3 5 4 7 8 7 7
3 2 4 4 4 8 4 4 4 5 5 4 4 4 5 9 9 8 9 9
3 4 2 4 3 2 2 3 3 3 5 4 4 4 7 7 8 9 8 7
2 4 4 3 3 3 2 2 5 3 5 4 4 4 7 7 8 9 8 7
4 2 3 3 2 2 3 4 3 5 4 3 3 4 9 9 7 9 9 8
2 3 4 3 3 3 3 4 4 5 4 5 5 7 9 8 9 7 9 7
4 3 4 3 3 3 3 4 3 4 5 3 5 8 8 9 8 9 8 8
4 2 3 3 4 3 3 3 5 4 5 4 7 7 9 9 9 9 9 9
4 3 2 4 4 5 5 5 5 3 3 3 9 8 8 7 8 9 8 7
2 4 4 2 2 3 5 4 5 3 3 7 8 8 7 7 9 7 8 9
3 2 4 3 4 5 4 4 3 5 4 9 8 9 8 7 8 8 8 7
```

Figure 10. Matrix of numbers representing the brightness (0..9) at each point in a 20x20 pixel array.

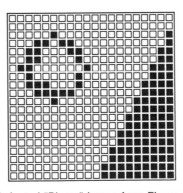

Figure 11. Discriminated "Binary" image from Figure 10, for pixels with brightness between 6 and 9.

Line	From	To	Line	From	To
2	7	7	9	5	6
3	5	9	9	8	9
4	4	6	10	7	7
4	9	10	12	19	20
5	4	4	13	17	20
5	9	10	14	15	20
6	3	3	15	13	20
6	10	11	16	11	20
7	4	4	17	9	20
7	10	10	18	7	20
8	4	5	19	5	20
8	10	10	20	3	20

Figure 12. Vertical chord table for the binary image in Figure 11.

more convenient) for the first and last pixels on each chord in the binary image. A table of these chords is quite compact, and contains all the information needed to define the shape of the profiles. Figure 12 shows a chord table for vertical lines comprising the image in Figure 11. This illustrates the amount of abstraction and compression that results from these encoding methods.

Once the image has been reduced to the perimeter points, the measurements proceed in the same way as for the semiautomatic case, except that quite complicated logic is required to decide which chords belong to the same feature outline. The method used requires that two chords touch each other at least corner to corner to be considered part of the same feature. Technically, this is referred to as 8–connectedness (each pixel can be connected to any of its eight nearest neighbors), as opposed to 4-connectedness (only the four pixels in the North/East/South/West directions are considered, not the diagonal neighbors). The latter allows for much faster computations in some of the image processing operations, and is satisfactory for some types of image recognition algorithms, but is inadequate for dealing with images in which the pixel size can be an appreciable amount of the feature, so that necks or protrusions may consist of only a few pixels. Most systems intended for image measurement use 8-connectedness for features (and hence they must use 4–connectedness for the complementary background or surroundings).

Because feature profiles may be concave, there may be branches that diverge or join in subsequent scan lines. This is usually handled in the programs, transparently to the user. Since the data are obtained in order of the scan lines, rather than all at once for each feature (as is the case for manually entered outlines), there may be several features present on one line. The chord table will contain all of their chords in scan order, so as they are unscrambled they must be assigned to individual features. It is also necessary to handle branching features. If the branches occur after a feature has been recognized, the bookkeeping is straightforward. But it is equally likely that the branches will initially be considered as separate features, and the information must be later merged when they are recognized to have joined together. Figure 13 illustrates some of these possibilities.

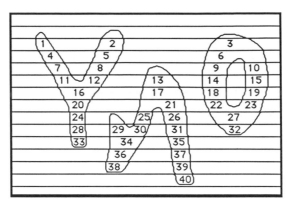

Figure 13. Complex features with the chords numbered in the order detected by scanning.

Binary image editing and modification

Because some images are less than perfect, the binary (discriminated) image just described may not represent the feature outlines exactly. For instance, there may be defects in the image due to the sample or illumination that need to be filled in. A good example of this is the secondary electron image (from an SEM) of spherical particles resting on a substrate. The sides of the particles will produce high yields of secondary electrons, and thus appear bright in the image. The substrate, being flat and featureless, is dark. But at the center of each particle there is also likely to be a dark spot, which shows up as a hole in the center of each circular feature in the binary image. An automatic routine to fill such holes before measurement can be applied to correct the image defect. This particular operation is carried out in software, using an algorithm that finds all pixels that are black and are not connected by other black pixels to an edge of the image, and makes them white. The details of the operation require no input from the user.

If the holes are not completely isolated from the outside, then it will usually require some user interaction to define the missing periphery so the figure can be closed in. This and some other operations call for image editing. The same devices (light pen, mouse, digitizer pad, etc.) as used to move the cursor on the screen for manual image measurement can be used to draw on the binary image, in either white (to complete features or join parts together) or in black (to separate touching features). Once this is done, the image is then ready for filling or measurement. Manual image editing is comparatively time consuming, and destroys some of the speed advantage of the automatic measurement method, so when too much editing is required it may be better to use the semi-automatic entry of features by visual discrimination and marking.

One good use of manual marking of the binary image is to mark the limits within which measurements are to be performed. For instance, if the illumination in a light microscope restricts the image to a more-or-less circular area in the center of the rectangular scanned video image, then drawing around the outline of the measurement area will allow the computer to know what the reference area for the measurement should be, in determining parameters like A_A and N_A. A particularly useful application of outlining is for the user to indicate the limits of a cell within which organelles are to be counted (and their size distribution determined). As discussed in chapter 2, the reference area for the measurements would be the area within the cell outline. And even for rectangular images, it may be useful to remove edge-touching features before measurement (since they cannot be measured in their entirety and will not be included in the feature-specific measurement results). This operation can be done automatically, since it is only necessary to start at each edge and follow joined chords, removing them from the binary image.

Another quick way to combine manual marking with automatic discrimination is to indicate features in the binary image that should be entirely rejected, or if it is more efficient, to mark only those that should be measured. Sometimes there are several visually distinguishable types of features that have the same brightness values. The automatic method will not be able to distinguish them during image acquisition, and both will appear in the binary image. However, a human operator may be able to instantly recognize those that are to be measured, and by touching them (or the ones to be removed) can very quickly obtain just the image to be measured.

In some cases the discrimination is based on size (very small features, in particular, may represent noise in the image), or shape. In this case, it is also possible to set limits in the programs which will ignore features which do not fall within those limits. However, this cannot be done until after the measurements have been performed. In general, the processing time to measure each feature in the image is proportional to the number of features (and it also increases with the number of parameters of interest, although some of them, such as length, breadth and angle, are all determined at one time using the same routines). Consequently, if the features of interest represent only a small part of the total number of features shown in the automatically discriminated binary image, it may be most efficient to select the ones to be measured manually.

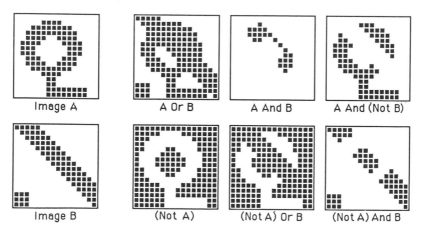

Figure 14 - Binary images and some of their logical combinations (pixel by pixel).

Several binary images may be combined if it will serve to isolate or demarcate feature outlines of interest. For example, images may be collected using different signals, such as different color illumination, different diffraction conditions in TEM, or different elemental X-ray maps in SEM (often combined with the backscattered electron image which has much better edge definition). Or, in fluorescence microscopy, the fluorescence image may be used to select features seen in better detail in the normal light microscope image. Each image must represent the same area of the specimen. Binary images from each source can be saved and then combined using various logical rules (discussed in Chapter 2), as shown in Figure 14. The resulting image can then be measured, combined with another image, or subjected to other editing operations.

Another use of image combinations is to select a subset of features to be measured or counted, for instance to restrict counting to features that lie within other features. A binary image can be formed for the larger ("parent") features, either by setting discriminators to a different value, by processing the grey image to obtain an edge (discussed below) and filling the outline in, or by manual delineation of an outline to be filled. This image can then be combined using an *AND* operation with the image of the interior ("descendant") features, to eliminate those which do not lie within the parent binary image.

As an example, consider the use of a biological stain such as Protein-A Gold, which deposits dark particles (of gold) to mark specific immune complexes for immunology or pathology studies (Figure 15). Many of these will appear within large grey secretory granules in tumor cells, examined in thin sections in the transmission electron microscope. Images of the granules (obtained by setting an appropriate grey level for the discriminators) can be filled to obtain a mask, which is then *AND*ed with the binary image of the darker gold particles. Then just the gold particles within the secretory granules are counted. The ratio of this number to the total count defines the specificity of the stain.

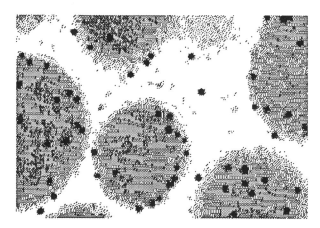

Figure 15. Protein-A-Gold stain deposits 15 nm gold particles (black) on immune complexes, mostly within secretory granules (grey). TEM image of pituitary tumor.

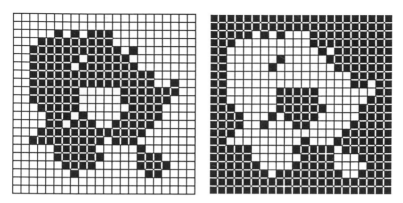

Figure 16. An example feature in a binary image, and its negative *(NOT)* or inverse.

Of all the image editing operations for binary images, probably the most used is etching and plating (also called erosion and dilation or dilatation). If a set of consistent rules are established for the fate of pixels on or adjacent to the edges of features, it is possible to smooth irregular outlines and in other similar ways improve the binary image. For instance, consider the image shown in Figure 17. The original outline has a very irregular profile, probably because there was little contrast between the feature and its surroundings and hence it was difficult to adjust the discriminator settings to obtain perfect separation. If the figure is used as-is, the area may well turn out to be OK, because the positive and negative errors around the edge will tend to cancel, but the perimeter will be much too large.

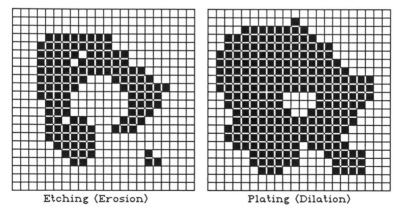

Etching (Erosion)   Plating (Dilation)

Figure 17. Etching and plating operations applied to the feature in Figure 16, using a test coefficient of 3 (etching removes pixels with at least three white neighbors, and plating fills in pixels with at least three black neighbors).

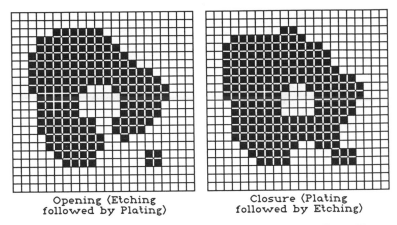

Opening (Etching
followed by Plating)

Closure (Plating
followed by Etching)

Figure 18. "Opening" and "Closure" applied to the feature in Figure 16.

A "plating" or "dilation" operation can be applied to this feature profile using a rule whereby each pixel which is currently "off" will be set to "on" if the number of its neighbors (the eight pixels which directly border it) that are "on" is at least equal to a minimum value. This will fill in the crevices as shown in Figure 17.

Similarly an "etching" or "erosion" operation would result if the rule were that any currently "on" pixel would be set to "off" if the number of its neighbors which were "off" exceeded a minimum value. This will erase minor protruberances, as shown. Both plating and etching are applied by totaling up the number of neighboring pixels that are on (for plating) or off (for etching). If the result (from 0 to 8) is greater than or equal to a comparison value, the pixel itself is set to the target condition (on for plating, off for etching). There is no universally optimal coefficient, since different constants produce slight geometric tendencies in the final outlines (emphasizing horizontal/vertical or diagonal edges).

Each of these operations changes the area of the feature, in proportion to the amount of perimeter that is present. But the sequential use of the two operations will leave the net feature area afterwards unchanged (on the average). Depending on the sequence, this operation may be called "opening" (erosion followed by dilation) or "closure" (the converse). The results are not identical, as shown in Figure 18, especially for small holes within the features (closing will tend to fill them in while opening will tend to join several into a single larger hole, or to the outside).

Serra (82) has discussed in detail the use of erosion and dilation on a hexagonal grid, as opposed to the square grid of pixels used here. His operations add or remove pixels based on he configuration as well and the number of neighboring pixels. In addition to modifying the shapes of features, these operations can be used to determine dimensions (by counting the number of cycles of erosion needed to entirely remove the feature), and to measure topological parameters such as the number of holes within objects.

Combination of a plated image with the original, using an *EXCLUSIVE-OR* function, will leave just the outline around the objects (etching, followed by *AND*ing with the original, produces a similar result in which the outline is the outermost layer of pixels, sometimes called the "custer" of the feature). These outlines are useful in several ways, including determining contiguity between phases as described in Chapter 3. Figure 19 shows an example, in which binary images are available for each of several phases in a microstructure (which may be distinguished by brightness, or color, or using X-ray maps obtained in the scanning electron microscope and made continuous using plating to fill the spaces between discrete dots). The outline around each phase is obtained using the method just described. Then the boundary images for each combination of different phases can be combined using the *AND* function. The fraction of the original pixel count for each phase's outline which survives the logical combination gives the percentage of the outlines shared by the two phases, which is the contiguity between the phases, or the probability that one phase will be found adjacent to the first.

A related technique is known as "skeletonizing," also known as a medial axis transform. This seeks to etch away all of each feature except for lines of pixels which represent the midlines and branching of the original shape. The number of separate branches (and/or the number of joints) and the length and orientation of segments will often reveal information about void growth and coalescence in ceramics, the orientation of fibers in paper, or patterns of growth in plant roots, to cite only a few examples.The topological interpretation of this information (see Chapter 7) is somewhat beyond the intended scope of this text.

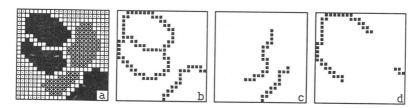

Figure 19. Binary images of three phases (superimposed in (a) for convenience in viewing). b) the outline around the black phase; c) and d) the common boundary with outlines around the other phases (39 and 48%, respectively, of the total boundary of the black phase).

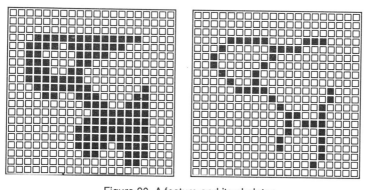

Figure 20. A feature and its skeleton.

Figure 20 shows an example of skeletonizing in which the result is achieved by repeatedly comparing the pattern of white neighbors around each white pixel to a set of selection rules (one for every possible pattern, 256 in all), removing ones that the rules reject, and then repeating until no changes take place. This removes outside pixels but preserves ones that have only two neighbors and thus must lie on the skeleton of the figure. The process has been likened to peeling an onion, or more descriptively to starting grass fires around an irregularly shaped field and letting them burn together. The final lines of ash then represent the skeleton of the original feature shape, and its branching pattern is clearly revealed. Separation of the branches can then optionally be performed (by erasing pixels that touch more than two other points) to study just the length and orientation of the branches, if desired.

Processing grey images

Before the binary image can be obtained (and perhaps edited) from the grey scale image, there are image processing operations that can be applied to improve the ability of the discrimination process to select the features of interest. There are several different levels of "effort" involved in these processes, some of which are only practical with substantial amounts of computing power or dedicated array processors. At the upper end, images can be transformed to the frequency domain (e.g. with the Fourier transform), and processed with filters before re-transformation to the spatial domain. These operations can sharpen images, remove artefacts (such as motion-induced blurs) due to the imaging system, and so forth. Other computationally intensive operations include "mapping," in which image distortion (for instance, if the surface is viewed at an inclined angle or must be rotated) is corrected. These operations are rarely needed for measurements of images that originate in the microscope, but for images from spacecraft (for example) they can be essential. Illustrations of these kinds of operations can be found in publications from NASA and the Jet Propulsion Laboratories (JPL). They require massive amounts of computing power, sometimes in the form of dedicated array processors.

Some of the other operations that JPL must apply to deal with images returned from remote space voyagers are appropriate for our more modest images and processing power, as well (Russ & Russ 84). These have principally to do with adjusting the range of contrast, so that the grey-scale separation of features of interest from their surroundings is maximized and made unequivocal. Some images are not uniformly illuminated (either because the light is uneven, or the camera has nonuniform response, or, for transmission images, the specimen thickness or average density varies from place to place). It is possible to "level" the background by subtracting (pixel by pixel) a background image obtained by recording the image from a suitable specimen (or in some cases no specimen at all). If no such image is available, it is often practical to obtain a representation of the background variation in brightness by fitting a polynomial to selected points in the image. For many purposes, the variation in illumination or thickness of sections can be well modelled by a simple second order polynomial in X and Y:

$$Brightness = A + BX + CY + DXY + EX^2 + FY^2$$

To determine the values of the constants $A...F$, it is necessary to have at least six points on the image that represent the typical background brightness; 20-30 are better, to permit fitting with plenty of degrees of freedom. These points are usually marked by the human operator, using a pointing device as discussed before. Once the polynomial image has been synthesized, it can be used just as a measured one would be. This is a much superior method to using an electronic filter to black out slowly changing brightness values, which was once a common circuit in video-scanned image analyzers. These operate only in one direction (the scan direction), and prevent making absolute brightness measurements to determine density. As computational power has increased, these analog tricks have become less important.

When the point-by-point subtraction of the background image from the measured one is carried out, it is possible to obtain difference values that are negative. Hence, it is usually necessary to add an offset value to keep the difference image (the "levelled" image) in the range of grey scale brightnesses that the system expects from a real image. The offset value most often used is the difference between the median brightnesses of the original and the background image (this is the value which has 50% of the pixels brighter and 50% darker, and is easily determined from the brightness histogram).

The illustration in Figure 21 shows an image from a transmission electron microscope (on a biological thin section), before and after levelling. Note that after levelling, the brightness values of similar structures are the same in different regions of the image, which is not true for the original. This permits the discrimination process to select features unequivocally. The same routines that subtract a background image can add or subtract any two (or more) images, in cases where that is appropriate (for instance to combine images obtained by different modes of operation). These operations are similar in intent to those carried out on binary images, as described above, and it is often better to work with the binary images when this type of combination is required.

The discrimination of different grey levels ultimately controls the ability to isolate features for measurement. However, for viewing purposes, the measured grey scale may not be optimum. Depending on the camera used, it may be linear with brightness, whereas photographs are logarithmic. Anyone with darkroom experience will appreciate that it can be very difficult to produce prints that range from black to white, while showing all of the features in the image. The same problem is present with digitized images, but there are more tricks available by computation than with photographic paper.

Figure 22 shows a representative recorded brightness spectrum from an image, and some ways that it can be translated to display brightness. If the entire range of brightness values is used, and linearly translated to display brightness, the result will be to represent the image as it was originally acquired (but with the number of brightness or color steps of which the display is capable - the human eye can only resolve about 15-20 grey steps). If some other transfer function is employed, it is possible to enhance some features of the original (always at the expense of others). For instance, the limits may be set to expand a small range of the total brightness, while clipping the remainder of the image to white or black.

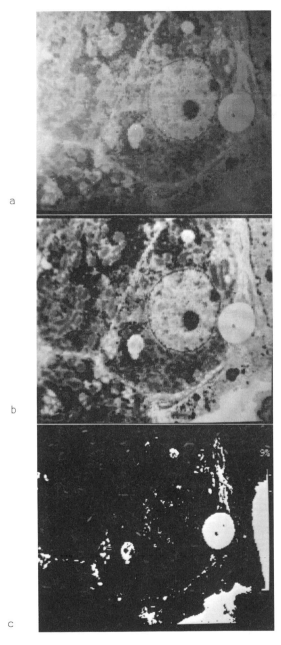

Figure 21. Transmission Electron Microscope image of biological thin section: a) original; b) after levelling background brightness variation; c) binary image of light grey regions selected from histogram.

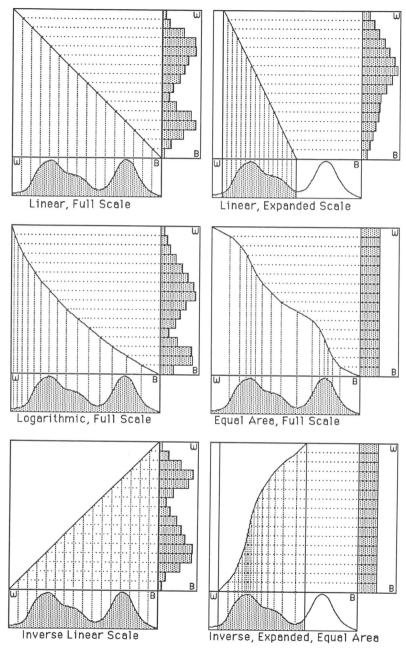

Linear, Full Scale

Linear, Expanded Scale

Logarithmic, Full Scale

Equal Area, Full Scale

Inverse Linear Scale

Inverse, Expanded, Equal Area

Figure 22. Transfer functions from the stored image brightness histogram to a display device's brightness scale.

Nonlinear transfer functions may also be used. A common one is logarithmic, much the same as a photographic print. This spreads out dark greys while compressing light ones. Another method, which cannot easily be duplicated photographically, is called "histogram equalization." This uses a transfer function that depends upon the individual spectrum. The goal is to produce an image with equal areas of each grey level, and this is done by dividing the brightness histogram into bands of different widths but equal areas. Then each band is displayed as one grey scale. The method is particularly good for expanding regions with sharp brightness gradients. Finally, negative transfer functions may be used (analogous to a photographic negative, reversing black and white). Any combination of these methods may be useful for particular images and features.

In the figure, the image's continuous brightness histogram is shown on the horizontal axis while the display device's brightness histogram is shown on the vertical axis, with discrete levels. Even more complicated transfer functions are possible, for instance with reversal to produce the equivalent of the photographic technique known as solarization, but these are rarely useful in an image measurement context.

If the display has color capability, it is further possible to assign a color to each displayed grey level. This can be used to further increase the visual separation between adjacent grey levels, since the human eye can only resolve fewer than 20 grey levels, but thousands of colors. However, the use of arbitrary color assignments can confuse the visual interpretation, by breaking up the continuity of features. It is generally better to use colors that vary gradually through changes in brightness, saturation or hue. For instance, the U. S. National Bureau of Standards (NBS) has recommended (Heinrich 84) a color scale that starts with dark blue, and proceeds through purple, violet, red, orange, yellow, green, cyan and then white, or one that follows the temperature curve. The brightness also increases gradually through the range. Other special scales may be used in particular situations.

These image processing operations do not improve the ability of the computer to separate features, but they may improve the ability of the operator to recognize or delineate features in the image. They also require a very minor amount of computing. Somewhat more complex are operations that sharpen edge contrast, extract texture in images, or smooth noisy images. These operations can all be carried out with the Fourier transform methods mentioned before, but equivalent results can also be obtained with an algorithm that can be applied with modest computational requirements.

The basic idea behind these operations is to compare the brightness of each pixel with those of its neighbors (8 nearest neighbors and perhaps the second neighbors as well), and from the average or differences, assign a new brightness to the pixel. The simplest type of operation to comprehend is to smooth out noise in the image by averaging the brightness values of the pixel and its neighbors. This is particularly appropriate for SEM images, because the backscattered electron image and especially the X-ray image can be very noisy due to the small number of electrons or X-rays at each point, and the correspondingly poor statistics of the image itself. Figure 23 shows an example in which an X-ray map is smoothed. Compare the typical X-ray "dot" map, which cannot be measured because it is not continuous

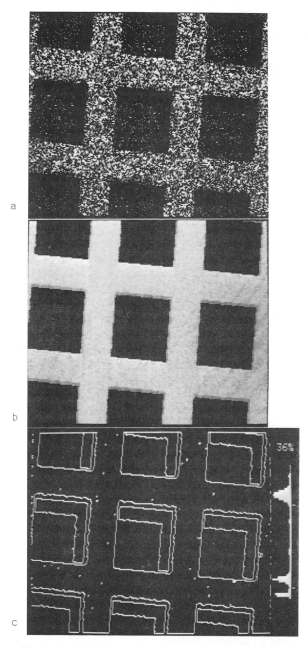

Figure 23. X-ray map from Scanning Electron Microscope: a) conventional "dot" map; b) grey scale map obtained by counting and smoothing; c) brightness histogram and contour lines.

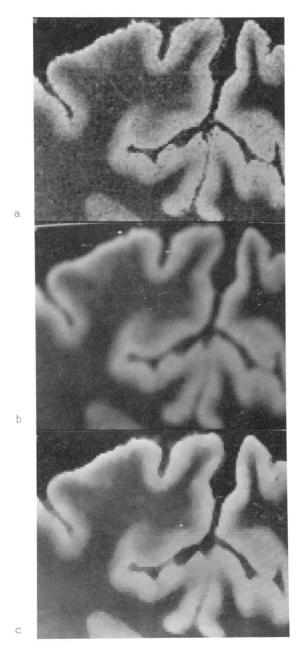

Figure 24. Enlarged portion of "noisy" autoradiograph of a cat brain: a) original;
b) smoothed with local operator; c) median filter applied.

and only with human interpretation can feature outlines be deduced, against the X-ray image acquired by slow scanning and counting X-rays at each point, and then smoothing that image.

Actually, smoothing an image using this method is less than ideal. Edges of features are blurred and reduced in magnitude, and may be shifted if the brightnesses on different sides of the boundary are quite different or vary in different directions. A better method is to use median filtering, in which each pixel and its neighbors are ranked in brightness order and the median value chosen to replace the original pixel. This does not shift or diffuse boundaries, and can be applied repeatedly if desired. It is particularly good for the "salt and pepper" type of statistical noise present in some electron microscope or X-ray images.

Figure 24 shows a portion (expanded to show individual pixels) of an autoradiograph of a section through a cat brain. In the original image, considerable pixel-to-pixel noise is evident. Smoothing the image reduces this noise markedly, but at the cost of edge definition, resolution of small features, and displacement of boundaries between areas of different overall brightness. The application of a median filter also smooths out the uniform areas, but does not cause the other side effects.

There are a variety of other operations which can be carried out using exactly the same type of operation as the smoothing described above, but with different rules for combining the pixel with its neighbors (Pratt, 78). Figure 25 shows some typical "kernels" or "operators." The numbers are used as multipliers for the brightness of each pixel. For instance, one Laplacian or differencing operator used simply multiplies the brightness of each pixel by eight and subtracts each of the neighbors from it. The result will be zero, or very close to zero, for regions of the image that are uniform in brightness, but large values will show up when edges or other sharp changes are encountered. Since the values may be either positive or negative, it is possible either to add an offset to the total to keep the result in range, or to use the absolute value.

Figure 26 shows the application of the smoothing (averaging) and Laplacian (differencing) operators to the same fragment of image shown in Figures 10, 11 and 12. To print the figure, a technique called "dithering" has been used, in which the number of dots printed on the page at each point is adjusted according to the point's brightness value. Since the original image had brightness values from 0 to 9, for each

Figure 25. Some representative grey scale image operators.

Original   Smoothed   Edge-Extract

Figure 26. Dithered representation of the image in Figure 10, and after applying smoothing with an averaging operator, and edge extraction with a differencing operator.

point on the image an array of 3x3 dots was used to represent the point, and from 0 to 9 of the dots were printed. This method has been used for most of the images in this book, and is a similar technique to that used in printing newspaper and magazine pictures, except that for the latter the dots are smaller and more numerous, and their size as well as number can be controlled.

Because the Laplacian is non-directional, it picks out features and feature edges regardless of orientation. This is particularly suitable for finding the edges of features that are sidelit or have complex variations in brightness that make them hard to discriminate from background by brightness alone. Examples are many particles viewed in SEM, or intracellular organelles viewed by transmitted light (especially with dark field illumination). Figure 27 shows the result of applying this kind of operation to an image from a videotape of bubbles in a liquid (whose size distribution was to be measured).

The new image formed by applying the edge extracting operator also has grey values for each pixel, proportional to the rate of change of brightness in the original image. The optimum binary image of the edges is formed by selecting approximately the brightest 15% for display. Alternately, the grey values of the edge image may be added back to the original image to brighten the edges while leaving the background for recognition of features; this is called edge enhancement.

Other operators such as those indicated in Figure 25 may be used to obtain derivatives that are directional, or to pick out texture running in a particular direction (for instance, to outline traces at a particular angle in integrated circuits). There are a wide variety of other operators published in the image processing literature for specific purposes, usually to extract some specific type of feature or detail from a raw image. Each of these operators has an equivalent Fourier transform operation, and the literature of image processing may discuss them in those terms. Many are 5x5 or 7x7 in size, rather than the 3x3 examples shown, but they work in the same way (although processing time increases geometrically as the size of the operator grows). In all cases, special treatment of the image edges is required.

The Laplacian operator is less than ideal as an edge-finding tool, as it has a stronger response to lines, line ends, and points than to edges. A variety of edge-finding algorithms have been devised. Some of these are computationally expensive and time consuming, and depend upon finding and following an edge, or segmenting the image into regions which are expanded from starting points until the entire image has been classified. These are more appropriate to the recognition of features in the kinds of images that are encountered in the field of machine vision and robotics,

a b

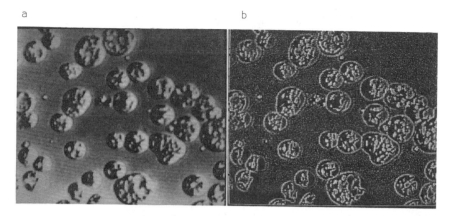

Figure 27. Bubbles rising in a liquid: a) original (note sidelighting); b) edges extracted with a differencing operator.

a b

Figure 30. Example of edge sharpening: a) original (alumina fracture viewed in SEM); b) increased edge contrast with Laplacian operator.

where they are most used. But there are simpler approaches that are appropriate to 2–D images as encountered in the microscope (and, to an extent, for satellite imagery).

The simplest of these methods (the Roberts' Cross) uses local operators to obtain directional derivatives in two orthogonal directions (this gives a measure of the "edge strength" in each direction), by subtracting the pixel brightnesses to get

$$E_1 = P(x, y) - P(x+1, y+1)$$

$$E_2 = P(x, y+1) - P(x+1, y)$$

These are then combined to get a non-directional edge strength (for a full discussion of this and other edge-detectors, with original references, see Rosenfeld & Kak, 1982). The combination should logically be the square root of the sum of squares of the individual directional values,

$$Edge\ strength = (E_1{}^2 + E_2{}^2)^{1/2}$$

but in the interests of reducing the amount of computing required, the simple sum of absolute values or the maximum value is often used instead. The Roberts' Cross operator is rather susceptible to very high frequency (pixel-to-pixel variation) noise in the image. A more widely used and closely related method is the Sobel transform, which uses the 8 nearest neighbors (numbered as in Figure 28) to obtain two derivatives:

$$E_x = N_2 + 2 \cdot N_3 + N_4 - N_6 - 2 \cdot N_7 - N_0$$

$$E_y = N_0 + 2 \cdot N_1 + N_2 - N_4 - 2 \cdot N_5 - N_6$$

The use of a 3x3 kernel makes this relatively efficient to compute, and more immune to high frequency noise than the Roberts' Cross. The methods for combining the two orthogonal partial first derivates to get the edge strength are the same, and in addition the local edge direction can be obtained as arctan (E_y/E_x).

Figure 29 shows an example of the application of the Sobel operator. The image (a light microscope image of a multiphase metal alloy) has several different regions, in this example easily distinguishable by their different brightness levels. A Laplacian operator does not extract all of the important edges, and increases the noise in the image substantially. The Sobel operator provides clean lines separating each of the phase areas, without adding noise.

Another edge-detector is the Kirsch transform, which obtains the derivative in each of eight directions (at each point in the image) and stores the maximum. The calculated pixel value is

```
0  1  2
7  P  3
6  5  4
```

Figure 28. Indices used to identify neighbors for pixel P.

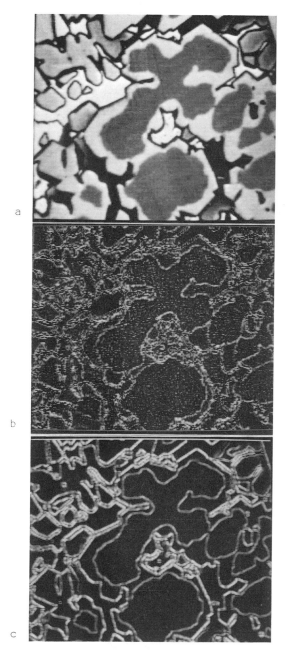

Figure 29. Light micrograph of a multiphase metal alloy: a) original; b) after application of a Laplacian (high pass) filter; c) after application of a Sobel operator.

$$Max\ (1,\ Max\ (E_0..E_7))$$

where

$$E_n = |\ 5 \cdot (N_{n-1} + N_n + N_{n+1}) - 3 \cdot (N_{n+2} + N_{n+3} +$$
$$N_{n+4} + N_{n+5} + N_{n+6})\ |$$

and the indices are understood modulo 8. These operators are generally preferred to simpler operations, such as imaging the difference between the mean (obtained by smoothing with a 3x3 kernel) and median, or applying a Laplacian to a smoothed image or to the logarithm of the brightness, since they are less sensitive to noise and do not displace edges. The resulting images are bright at edges, and discriminators can be set to obtain a binary image of the edges, which are in turn filled in to obtain features for measurement (etching or skeletonization may be needed first to remove small bits of noise that do not come from edges). When incomplete edges are obtained, it is possible to use a Hough transform in which the local edge direction information is used. Each point "votes" in proportion to its edge strength, and inversely to its distance; the weighted mean of the orientations is used to extend the edge. This takes rather more computing than most of the other methods that have been discussed.

Sometimes, grey scale image operators are simply used to increase local contrast changes at edges in images, as shown in Figure 30. In this case, an edge extraction operator was applied that had its center value or weight increased. This has the effect of adding some proportion of the edge information back to the original image, so that the original grey scale information is still evident, but somewhat compressed in grey range, and the edges are enhanced. Because of the way the human visual system works, this has the effect of sharpening the appearance of images, and may be useful whether or not the images are intended for measurement, or to assist visual discrimination of features for manual demarcation and semi-automatic measurement.

Chapter 6

Data Interpretation

Differences from manual methods

With the automatic and semi-automatic image measurement methods, the parameters that are directly determined (area, perimeter, Feret's diameters, length, breadth, angle, location of center of gravity, etc., as described in the preceding chapter) can be used directly to characterize many structures. Particularly in the case of sections through matrices, in which profiles outline different structures or phases, the analysis is much like that we have seen before with manual methods. For example, the volume fraction V_V is simply equal to A_A, the total area of features divided by the reference area of the image. This can determined quickly from the sum of the chord lengths divided by the known dimensions of the image. Likewise, the total perimeter is just the boundary length per unit area, from which the surface area per unit volume is obtained as $S_V = 4/\pi\,B_A$ (the reason for the $4/\pi$ is that in general the boundaries are not perpendicular to the surface being examined).

The most common measurement performed by manual methods is the counting of intercepts along random test lines, to determine the mean intercept length. For automatic measurements, the lengths of chords in the chord table can be used instead. The total length of the the chords is known (it was just used to obtain the area fraction of the features in the binary image), and the total number of chords can be determined simply from the length of the chord table. The ratio of these gives the average chord length, which is just the desired mean intercept length. From this mean intercept length, the grain size number and other parameters can be obtained.

Size distributions were handled in the manual methods by obtaining histograms of intercept length or profile size. Since the manual methods can obtain profile size directly, this is generally used. The "equivalent circle diameter" is often obtained from the measured feature area, as the size of a circle which would have the same area as the profile in the binary image:

$$d = \sqrt{\{4\,A\,/\,\pi\}}$$

These diameters can then be used to determine the size distribution of generating spheres, as in the manual operation described in Chapter 4. It is particularly attractive to utilize the model of ellipsoids of revolution, because the length and breadth of the individual profiles can be readily determined as the greatest and least Feret's diameters at a series of angles (as discussed in the preceding chapter). From the length and breadth, the Dehoff method for determining N_V and the size distribution of the 3-D features (assuming that they are either oblate or prolate

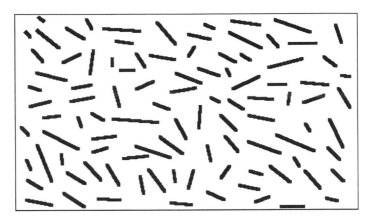

Figure 1. Binary image showing magnetic iron oxide particles (from a transmission electron micrograph).

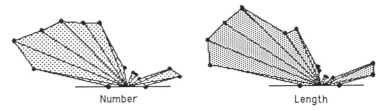

Number   Length

Figure 2. Plots of number and cumulative length of particles in Figure 1 versus the angle of the longest dimension. The plotted quantity is represented by the length of the vector drawn at each angle.

spheroids) follows directly, using a matrix of coefficients that can be stored in the computer.

In addition to histograms of number of features versus size, which were considered under manual methods, there are some other interesting plots that are readily obtained with automatic (computerized) instruments. For instance, we previously mentioned examples of plotting the number of features or total length of features as a function of angle, to characterize the degree of orientation. Figures 1 and 2 show an example of this, in which the binary image is obtained by discrimination of a video image of dispersed magnetic oxide particles (as used on computer disks). The plot of number of features versus orientation (in 16 equal angle steps) covers only the range of 0-180 degrees (the range from 180-360 is redundant), and is shown in a different format than most distribution plots because the "compass rose" pattern makes it easier to interpret the data. The angle is that of the longest Feret's diameter, which lies along a diagonal of the rectangle; however, that is close enough to the longest side of these features to show the trend.It is clear from Figure 2 that these particles are not randomly oriented (many more lie at angles between 110 and 170 degrees than in other directions). The second plot shows the cumulative length of features sorted into each bin by their angle. Any difference between this and the number plot would show the tendency for alignment to vary with length of the particles.

There are other plots that may likewise be interpreted to study the physical distribution of particles, organelles, etc. in a matrix or structure. Plots of number, size or angle versus location may reveal nonuniformity (for instance near surfaces where composition of metal alloys can vary, the size of grains or precipitates may vary, or the orientation of microfibrils may change near membranes to which they attach).

Projected images

It is for the case of projected images of dispersed features in a section or particles on a substrate that the parameters provided by automatic measurement allow some new possibilities in interpretation. Because the greatest use of these tools has been to characterize particulates, either raw materials (which are encountered in many industrial processes), fragments of broken or ground materials, air or water pollutants collected on suitable filter substrates, residues lifted from surfaces with adhesives (including extraction replicas used in the electron microscope), and so on, we will refer to the features as particles. The reader should understand that other samples are also appropriate to this treatment, including cell organelles viewed in thick sections (so the entire feature outline is present), astronomical photographs of planetoids, and much more.

The key is that the projected feature outlines show directly representative sizes, such as caliper diameters. If it is permissible to assume that the features are of some predictable shape, such as flat layers of constant thickness (for instance, particles of clay minerals, which tend to lie flat with proper preparation, or pieces of leaves recovered from a cow's stomach to assess breakup during chewing), or cylinders or ellipsoids of revolution lying on their sides (a general model for many real shapes), the volume can be estimated from the outline. This gives rise to a set of derived parameters that can be calculated, for each feature, from the measured ones (Russ & Stewart 83). For instance:

VOLUME may be taken as the volume of an ellipsoid of revolution with its major axis equal to the length of the outline, and minor axis equal to the breadth (other estimators for the major and minor axes will be presented shortly). This assumes that the feature is lying on its side, which (depending on the specimen preparation technique) is usually more likely than finding it standing on end. Other specific shapes, such as cylinders, may also be appropriate in some cases. For the case of a prolate ellipsoid, the volume is simply:

$$Volume = \pi/6 \cdot (length) \cdot (breadth)^2$$

While this estimate of volume is very commonly used, and is adaptable to many different types of real features, if it is applied to features that are very different from the assumed model (such as flat platelets, pyramids, etc. as shown in Figure 4), quite large errors in volume may result. This will be discussed shortly.

Some other derived parameters are less immediately obvious than the volume, but become quite useful in plotting size distributions, or as intermediate steps to calculate other parameters. For instance:

EQUIVALENT SPHERICAL DIAMETER is defined as the diameter of a sphere whose volume would be equal to the volume as calculated above. It is often used to characterize a linear size dimension, for purposes of creating size histograms (as we shall discuss shortly), without regard to shape, or to plot things such as shape or orientation versus size. Since the volume as defined above is obtained from the length and breadth, the Equivalent Spherical Diameter can be gotten directly as

$$Diam. = \{\, 6/\pi\, Volume\,\}^{1/3} = length^{1/3} \cdot breadth^{\,2/3}$$

There are other equivalent size parameters which are sometimes used instead, depending on the expected nature of the features. For instance, the equivalent circular diameter (the diameter of a circle with the same area as the area of the measured outline) would be appropriate for the case of flat objects, as well as for sections through features in a matrix as described before.

FORM FACTOR is often used to describe the shape of features. It is usually defined as

$$Form\ factor = 4\pi\, Area\,/\,Perimeter^{\,2}$$

which will be 1.0 for a perfect circle and will decrease in magnitude as the outline becomes more irregular (a square has a form factor of 0.785, as shown in Figure 3, while other shapes may have much lower values).

The form factor is often used in calculating other parameters, but is also useful by itself. For instance, if plotted against size, a trend of decreasing form factor with increasing size (e.g. equivalent spherical diameter) would suggest that large particles are more irregular than small ones, which could reflect the agglomeration of small particles to form large ones. Conversely, a trend of decreasing form factor with decreasing size could result if small irregular particles were produced by fragmentation of large smooth ones.

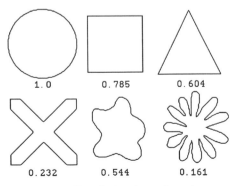

Figure 3 . Form factors for various shapes.

Figure 4. Strongly polygonal, preferentially oriented, or locally concave shapes whose surface area cannot be accurately estimated using the model incorporating form factor, as described in the text.

ASPECT RATIO is simply the ratio of length to breadth, and is also used as a descriptor of feature shape in some cases.

SURFACE AREA is a very important characteristic of particles, whether they are catalyst powders (rates of reaction are proportional to surface area), ceramic raw materials (the ability to press the powders together and sinter them will depend on the amount of surface area), or organelles (diffusion of molecules is often related to the amount of membrane surface area). Unfortunately, surface area is not directly measured by these techniques. If we were to use the surface area of the ellipsoid of revolution described above for volume, we would clearly underestimate the surface area (because small irregularities or departures from a smooth ellipse would add to the surface area much more than they would change the volume). One common method used is to divide the surface area of the ellipsoid by the form factor, as discussed below.

Strongly polygonal shapes will show considerable bias against this simple model. For instance, if the projected shape is a square of side length 1, the ellipsoid of revolution fit to the length (1.414) and breadth (1.0) will have a volume of 0.7404 cubic units and will give an estimated particle surface area of 4.865 square units. However, if the actual shape of the particle was a cube, the true volume would be 1 and the surface area 6. If the actual shape were a pyramid of altitude 1, the volume would be 0.333 and the surface area 3.236. Figure 4 shows some of these problem shapes.

And, of course, if the particle has deep holes which are not revealed on the outline (e.g. if the particle is seriously concave), then the actual volume will be reduced, and more importantly the surface area may be substantially increased, compared to the estimated value. Strongly preferred orientation, so that the height cannot be estimated from the breadth of the projected feature, is also a problem. Still, the simple estimate given by the equation represents a useful average for many typical particles which are able to be approximated as ellipsoids of revolution.

An even simpler estimate of surface area for convex particles can be obtained from the relationship of Cauchy (1832). If A_{mean} is the average projected area from many identical particles viewed in all orientations, then the surface area is just

$$S = 4 \cdot A_{mean}$$

Many other derived variables can be defined in terms of the measured parameters. Some will be important for an entire class of specimens, while others may have more limited use. For instance, in the textile industry, a common descriptor of the shape of fiber cross-sections is called the "modification ratio," the ratio of the diameter of the an inscribed circle to the diameter of a circumscribed circle. This is clearly a holdover from the days of manual measurement, since it was comparatively easy to lay a circle template over the image to determine the two circle diameters. Although the more complete measurements possible with modern automatic systems could probably be used to determine other combinations of parameters that would more directly report the state of wear in the dies through which the fibers are drawn, the modification ratio is desired for consistency with long-standing tradition and experience. For fibers with a given shape (the number of points in the star, triangle or other cross section used to stiffen the fiber while reducing its mass), it is possible to obtain a function of area and perimeter which correlates very closely to the modification ratio, and can be used for this purpose.

Another example of a very simple derived parameter arises in the measurement of rectangular features. It was pointed out before that the length is actually determined as the longest Feret's diameter, and hence is the diagonal of the rectangle. If the length as it is normally defined is required, it can be obtained as Area/Breadth.

Accuracy of derived parameters

In order to properly use the derived parameters introduced above, it in important to understand the assumptions they represent and their expected errors. First, consider the parameters that can be directly measured on the image: Area, Perimeter, Length (longest Feret's diameter), Breadth (shortest Feret's diameter), Angle (orientation of the Length direction), Position (X and Y location of the centroid of the shadow), and perhaps Brightness or Density if the particles are not totally opaque. From these, we will want to estimate such things as the particle volume, surface area, and some descriptive shape factor(s).

As usual, we need some conceptual model for the three-dimensional shape. In some cases the shape of the particles will be known from other means, in which case an appropriate model should be used (eg. long thin cylinders for fibers, thin disks for platelets, etc.). The local thickness can sometimes be estimated from the image density to truly reveal the 3-D shape. However, in the absence of other information, a common assumed model is the prolate ellipsoid of revolution, defined

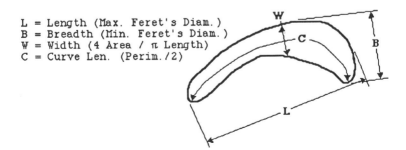

```
L = Length (Max. Feret's Diam.)
B = Breadth (Min. Feret's Diam.)
W = Width (4 Area / π Length)
C = Curve Len. (Perim./2)
```

Figure 5. Feature dimensions discussed in text.

by its major and minor axes. We saw in Chapter 4 that this is often a useful model for many real objects, and is also easy to use in calculations.

The ellipsoid axes can be estimated in several different ways from the measured parameters. For instance, the length (maximum Feret's diameter) can be taken as the major axis, and the breadth (minimum Feret's diameter) as the minor axis. However, as indicated in Figure 5, if the projected image has any significant concavity in its shape, it might be better to use the width for the minor axis (width is here defined as *4·Area / π·Length*). This would more closely conform to the actual particle. Furthermore, if the feature is relatively long compared to its width, it may be preferable to estimate the major axis by the curve length (here defined as half the perimeter of the projected image, which is obviously most appropriate for fibers of negligible width).

There are other combinations of parameters which can also be used, and are most appropriate for particular classes of shapes. But if we use these definitions, there are at least three ways to estimate the axes *A* and *B* of the ellipsoid, from which the volume may be obtained as $\pi/6 \cdot A \cdot B^2$. We will refer to the combinations as *(L,B)* for the length and breadth, *(L,W)* for the length and width, and *(C,W)* for the curve length and width.

If known three-dimensional shapes are rotated in space to their extreme positions (at which they produce their maximum and minimum projected or shadow images), and then the area, perimeter, length, and breadth of the shadow are determined, the volume can be estimated using an ellipsoid generated from all three of the models. In Figures 6-8 this has been done for several regular polyhedra (tetrahedron, cube, octahedron, and icosahedron), as well as cylinders and ellipsoids of revolution (with length to diameter ratios of 0.2:1, 0.5:1, 1:1, 2:1, and 5:1).

Some trends are immediately obvious, and indeed should have been expected. First, very angular shapes like the tetrahedron are not really very well

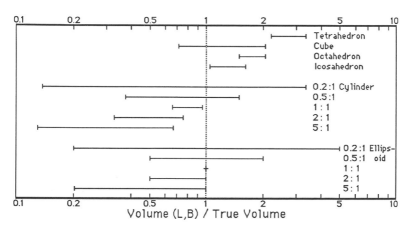

Figure 6. Range of errors for volume estimated from ellipsoid of revolution with major axis equal to length and minor axis equal to breadth, for a variety of geometric shapes.

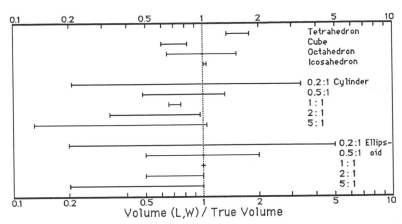

Figure 7. Range of errors for volume estimated from ellipsoid of revolution with major axis equal to length and minor axis equal to width, for a variety of geometric shapes.

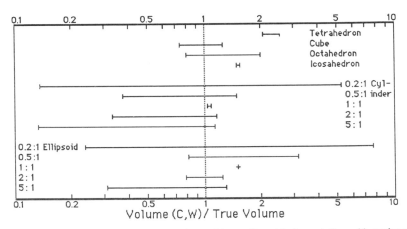

Figure 8. Range of errors for volume estimated from ellipsoid of revolution with major axis equal to curve length and minor axis equal to width, for a variety of geometric shapes.

modelled by a smooth ellipsoid of revolution, while the more regular shapes generally show better agreement between estimated and actual volume. The trends suggest that the *(L,W)* model is perhaps the most generally reliable, while the *(C,W)* model is superior for long skinny shapes (which it should be, as these are the shapes upon which the model is based). Even for such expectedly terrible models as approximating an oblate shape such as a short ellipsoid or cylinder, the results may be useful (in this case a model based on an oblate ellipsoid would far superior, if the object was known to have that shape).

Second, there is often a considerable range to the estimated volume, depending on the orientation of the solid. This is especially pronounced for shapes like the cylinders, for which the projected (shadow) images are very different in the extreme positions (one a circle and the other a rectangle). In an actual random sample with all orientations present, these will average out considerably. Notice that the *(L,B)* and *(L,W)* models are identical in their predictions for ellipsoids of revolution, and are exactly correct when the shadow image shows the maximum aspect ratio of the generating shape.

This points up the fact that we do not on the average see the maximum projection of the generating shape. For three-dimensional ellipsoids of revolution with axes *A* and *B*, the "Aspect Ratio" is defined as *A/B*. The observed aspect ratio *L/W* in the two-dimensional projection is usually less than this. If all orientations are equally likely, an integration can be performed to determine the mean relationship between the actual and observed aspect ratio. This gives approximately

$$L \,/\, W = 1 \,+\, (\pi \,/\, 4) \cdot (A - B) \,/\, B$$

which behaves as we would expect in the limits. As *A* approaches *B* (the case of a sphere), *L/W* approaches a value of 1. Likewise, for *A >> B*, this approaches *A/B·(π/4)*, or the observed projected length approaches $\pi/4$ times the actual length, which is the expected projected length of a needle in three dimensions (the generalized Buffon needle problem mentioned before). Solving for the aspect ratio of the 3-D feature gives

$$(A/B) = 1 \,+\, (4/\pi) \cdot (L/W - 1)$$

which we can use to estimate the aspect ratio of each feature from its projected image.

Aspect ratio is a crude estimator of shape, although in some cases, plots of aspect ratio versus size or other parameter may show important trends in the material being examined. A more widely used derived parameter is the "form factor" described above

$$4 \,\pi \, Area \,/\, Perimeter^{\,2}$$

(sometimes, the inverse of the form factor as defined here will be encountered instead). This dimensionless parameter is 1.0 for a perfect circle, and decreases for less regular shapes; thus we can use form factor to describe the shape of 2-D outlines, whether obtained by sectioning through features, or viewing the projected image. Correlation plots of form factor against size, orientation, etc. have already been mentioned, and an example is shown later in this chapter.

For the present, we want the form factor in order to attempt to compensate for the fact that the ellipsoid of revolution that we have chosen as a 3-D model for particles has a minimum surface area for its size (volume and aspect ratio), and real shapes which are "rougher" will have more surface area than would be estimated using the ellipsoid alone. By dividing by the form factor, a larger surface area is obtained. Using the same shapes as described above for the volume comparison, we can compare the estimated surface area to the actual, using the *(L,B)*, *(L,W)*, and *(C,W)* ellipsoid models. The formula for the surface area of a prolate ellipsoid with major and minor axes *A* and *B* is given by

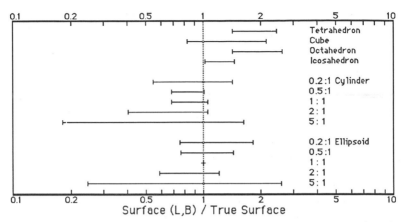

Figure 9. Range of errors for surface area for an ellipsoid of revolution with major axis equal to length and minor axis equal to width, divided by the form factor of the projected shape.

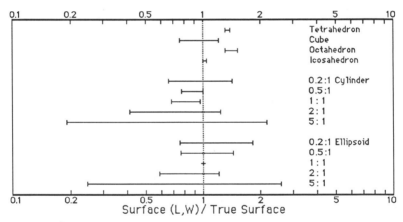

Figure 10. Range of errors for surface area for an ellipsoid of revolution with major axis equal to length and minor axis equal to breadth, divided by the form factor of the projected shape.

$$(\pi / 2)\, (B^2 + A\, B\, sin^{-1}\, (e)\, /\, e)$$

where the eccentricity is

$$e = (A^2 - B^2)^{1/2} /\, A$$

The plots in Figures 9-11 show the results. Surprisingly, even with the rather large range of shapes, the results for surface area are better than the earlier ones for volume. Again, the *(L,W)* model is generally superior to the others, while the *(C,W)* model is very good for long skinny shapes, and poor for oblate or disk shapes. It is interesting to note that the estimated surface area of ellipsoids is generally too high. This is because the act of dividing by the form factor to increase surface area for an irregular shape is exactly the wrong thing to do when the shape is actually smooth.

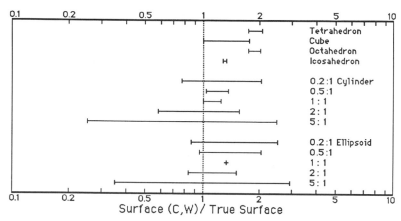

Figure 11. Range of errors for surface area for an ellipsoid of revolution with major axis equal to curve length and minor axis equal to width, divided by the form factor of the projected shape.

The form factor for the projected ellipse is less than 1.0, and erroneously increases the area. If it was known that the generating shapes were ellipsoids, this overcorrection could be eliminated. In general, most particulates do have some irregularities and roughness, and so the form factor correction is usually appropriate.

Use of the derived parameters

Once they have been defined, the various derived parameters can be calculated for each feature and then used in all of the usual statistical ways. This is a good example of the difference that computerization of the instrument makes, as it would be unrealistic to carry out such calculations for a statistically useful number of particles, by hand.

In some situations, the mean value of the form factor or volume may be of some interest. But it is distribution functions that make the best use of this type data, and they are rich with stereological meaning. We saw before that histograms showing the frequency distribution of feature size can be obtained, with either a linear or logarithmic size scale. Normally the scale will be set up with counting bins arranged with some meaningful size step and limits. In the case of a logarithmic scale, it is the ratio of sizes from bin to bin that is the step size, and values such as 1.1 or 1.2 (10% or 20% increase from one bin to the next), 1.414 (square root of 2, so that each bin contains features with twice the area as the last), and so on are common.

The number of bins that can be used increases as the number of observations does, because of the requirement for enough counts in each bin to give reasonable statistical precision for a smooth histogram. As discussed in Chapter 1, the standard deviation for counting is just the square root of the number of counts. At least thousands if not tens of thousands of features must be counted to produce acceptable counting precision in the sparsely populated bins of a distribution, which may contain only a few percent of the total observations. A sufficient number of observations may be obtained by combining the results from many fields or images, all at the same magnification (see the comment in chapter 4 about combining distributions of large and small features measured at different magnifications). Log distributions are generally better than linear ones in terms of the statistics of the sparsely populated upper bins (which dominate volume and surface area plots).

The precision problem is aggravated by subtraction. If the size distribution of one material must be determined from a "pure" sample and then subtracted from another distribution containing the original material plus a small admixture of another type or size range, the propagation of errors is severe for the minor amount determined by difference. For instance, if in a given bin, the number of counts of pure material A is N_{Ai}, and that for a composite material B is N_{Bi} (each with a standard deviation of the square root of the number of counts), then the difference is $N_{Bi} - N_{Ai}$ but the standard deviation is the square root of the sum of the two, or $(N_{Ai} + N_{Bi})$. If $N_{Ai} = 100$ and $N_{Bi} = 110$, the net difference is 10 counts and the standard deviation is over 14, so it would be impossible to decide with confidence if the second component were even present. With $N_{Ai} = 10,000$ and $N_{Bi} = 11,000$ the difference is 1000 and the standard deviation is only about 145, so the relative precision is much better. Of course, this requires performing 100 times as many measurements!

Instead of plotting the number of features (or the effective number, if the method described in Chapter 4 for correcting for the finite size of the image area and its effect of limiting the ability to measure large features is employed), it is also possible to plot some other parameter on the vertical axis. This can be summed for all features which were sorted into the bin or category based on the size, or other sorting variable. Some simple examples have already been given, such as the total length of features sorted by their angle, to demonstrate the amount of preferred orientation. The derived parameters can also be used in this way.

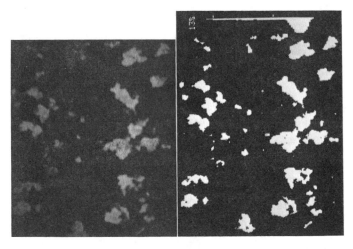

Figure 12. Paint pigment particles viewed in SEM, and binary image used for measurement in Figures 13, 14 and 17.

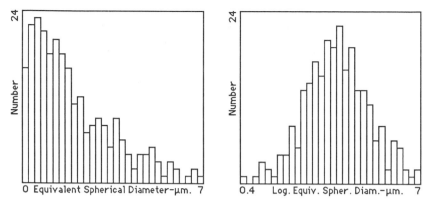

Figure 13. Size distribution plots for paint pigment particles, using the equivalent spherical diameter as a measure of size.

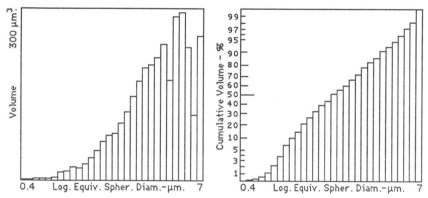

Figure 14. Plots of volume and cumulative volume vs. log size, for the features shown in Figure 13.

For instance, plots of the total volume or surface area of particles sorted by a measure of size (e.g. the equivalent spherical diameter) show graphically the importance of the larger features. Figure 12 shows an image of paint pigment particles. Measurements of discriminated binary images from many viewing fields were combined to provide data on 264 particles. Figure 13 shows plots of the number of particles, sorted by the equivalent spherical diameter. Both linear and logarithmic sorting axes are shown, and it is immediately clear that in this case the log plot shows a much more symmetrical and interpretable distribution (this is often the case for real particulates produced by physical grinding or other similar operations).

While the most probable particle size is rather small, it is the few larger particles that contribute most of the volume. The importance of the large particles is readily shown by the plots in Figure 14. Here, the volumes (calculated from the ellipsoid of revolution fitted to length and breadth) for all features in each size class (based on the logarithm of the equivalent spherical diameter) are totalled, and plotted. The second plot shows the cumulative volume (for the class itself and all smaller classes, as would for instance be obtained in sieving experiments from the amount of material that passed through a given size screen). Furthermore, the vertical axis is a probability scale, as was described in the first chapter. The result is a nearly straight line fit (indicating a log-normal distribution).

The deviation from linearity that causes the lower size classes to fall below the line may be due to undercounting of small features, which can hide next to or under larger ones in the projected image. Small features may also be lost on the binary (discriminated) image because of low contrast, or because they are erased in the process of smoothing or etching and plating. In an image of a polished surface, small features are also lost because they are more likely to be pulled out in the polishing operation. It is this deviation that represents the undercount problem mentioned before, and provides the justification for the method presented in Chapter 4 for completing distributions by extrapolating the contents of small size classes.

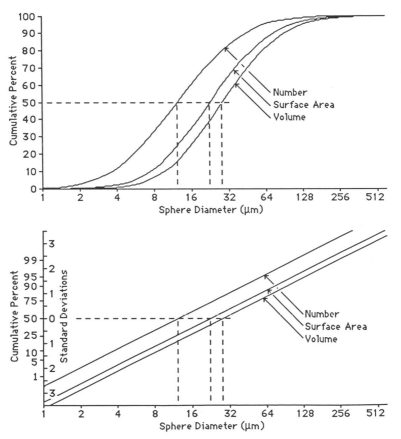

Figure 15. Plots of cumulative volume, surface area and particle count for a log-normal distribution of spheres.

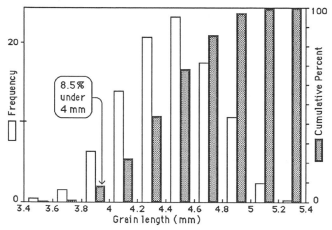

Figure 16. Size distribution and cumulative plot of 500 rice grains sorted by length, from a series of live video images.

Plots of cumulative volume produce straight lines in log-normal plots on probability paper. So will plots for surface area or number of features, except that they are displaced relative to the plot for volume. The point at which each of these curves crosses 50% (the midpoint of the curve) gives the mean size (the geometric mean in the case of a logarithmic size scale) of the particles. This means that there are three different mean sizes: one that corresponds to the mean size based on number of particles, one for mean size based on the volume, and one based on surface area. The appropriate mean size should be selected in describing particle behavior that depends on volume or surface area (for instance, any property having to do with surface reactions will normally be best represented by a mean size based on the cumulative surface area plot).

This effect is shown graphically in Figure 15. Cumulative plots of number, surface area and volume for an ideal log-normal distribution of spherical particles are shown both with linear and probability scales on the vertical axis. Note that the 50% point for each parameter corresponds to a different diameter (more than a factor of two difference). The "mean" size based on volume, which would be relevant if the particles' mass controlled their behavior, is much larger than the "mean" size based simply on diameter, which is the size parameter usually reported by image measurement methods.

The mean values from these determinations can be compared between several samples, or between different fractions of one sample sorted by some other variable. Another common use for cumulative plots is in quality control applications. Specifications for products often set a minimum or maximum fraction that can be below or above a certain limit. Figure 16 shows an example, in which the original images were obtained by scattering long-grained rice on a dark background (velvet cloth was used) beneath a TV camera. The requirement for long-grained rice is that no more than 10% of the grains have a length less than 4 mm. Scans of about 500 grains were measured, in several images, and automatically discriminated. Grains which touched were rejected because the formfactor was too low. The size distribution and cumulative plot of the remaining grains were sorted by length. The fraction under 4 mm. (8.5% in the example) is immediately evident.

A similar application in another grain involves testing wheat for a permitted maximum amount of soft wheat in the more costly hard (winter) wheat. In this case the rounder shapes of the soft wheat are distinguished by the aspect ratio (length/breadth). Plots using this derived parameter as the sorting variable give the desired result.

This illustrates that either measured or derived parameters, in any combination, can be used for distribution plots. For instance, features can be grouped based on angle (orientation) or form factor, and the plot of total size in each group compared using an analysis of variance test. This would reveal whether the size varied with these parameters.

Scatter plots of the data will often show the same thing. Figure 17 shows a plot of the form factor versus size for the paint pigment in Figure 12. The drop in the form factor for larger particles is apparent. In this case it is due to agglomeration of small particles to form the larger ones. However, as it is the agglomerated particles that control the behavior of the paint, it is the mean size of the particles which has been determined that is important in this application. Note in this figure that while the correlation coefficient is quite high, and definitely shows a significant correlation between size and shape, there is a definite curvature in the trend shown by the data.

In addition to scatter plots, it is sometimes useful to collect data in a two-way or cross-tabulation plot, in which sorting bins for two variables are set up, and features are counted in each bin. An example of this type of plot is shown in Figure 18. These displays show where the points lie better than the scatter plot does, because the human eye does not measure dot density very well, and when there are a great many points on a scatter plot, some will overlap.

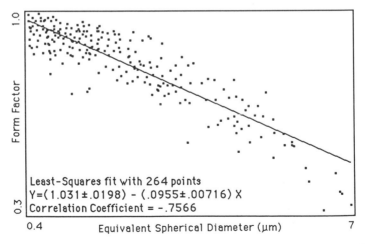

Figure 17. Scatter plot of form factor vs. size for the particles in Figures 12-14.

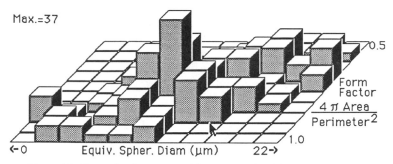

Figure 18. Example of a two-way histogram using bins to sort features by size and form factor.

Density and Roughness

All of the measurement parameters for objects described so far, can be obtained by measurement and/or calculation from the binary, or discriminated image. Sometimes it is useful to look back at the original brightness values for pixels which lie within the features in the binary image. Note that these pixels do not necessarily have the same, or even a small range of brightness values. The feature outlines may have been obtained from various processing steps, such as edge-finding algorithms described in Chapter 5 (or even from manual delineation). But by using the pixels in the binary image as a mask, it is usually possible to read through the original grey-scale image to find the brightness values in the original, unprocessed image.

Several properties can be determined from these values. The most obvious is density. In a transmitted light or electron image, the brightness measured by a linear device (most video cameras) is related logarithmically to the density of the sample. Some recording devices are themselved logarithmic (film, for example), and provide the density directly. In either case, calibration of the measured mean brightness of the image against known density standards is straightforward provided the illumination remains constant (and uniform) and the gain or aperture of the camera is not changed. Depending on the application, the density may also be converted to other, derived units such as concentration of a staining element, dosage of radioisotope labelled compounds in autoradiography, and so forth. In fluorescence microscopy, the mean feature brightness can also be measured and converted to meaningful units.

Besides the mean brightness of pixels within features, we can measure the variation in brightness as a function of distance. For instance, the brightness in a normal SEM secondary electron image is a sensitive indicator of surface orientation, and the variation may characterize the degree of roughness of a surface. Haralick (73) has presented an extensive set of parameters that can be determined from pixel brightness differences to characterize the texture of surfaces. These include means, variance, entropy (based on the logarithm of mean difference over various distances), and moments of the brightness variation versus distance distribution. The parameters serve as identifiers of particular terrain in satellite photos. As originally presented, these determined texture information from the brightness differences between adjacent pixels, in each 45 degree direction. When the orientation of features is

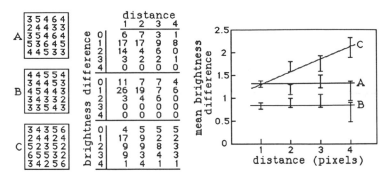

Figure 19. Three image fragments with their brightness values and plots of brightness differences as a function of distance.

randomized, it is equally useful and computationally more efficient to use differences only in orthogonal directions (which also avoids the problem that pixels are 40% farther apart along the 45 degree directions than along the X- and Y-directions).

Figure 19 shows examples of different fragments of grey scale images (with brightness values reduced to a single digit for clarity). For each, the table shows the number of pixel pairs with various brightness differences as a function of distance. From this frequency information, the mean and standard deviation of each column in the table was used to construct the plots shown. Note that for image A the mean variation in brightness between adjacent pixels, or for that matter between pixels any distance apart within the feature, is greater than for image B. On the other hand, for both A and B there is no significant trend of brightness difference with distance, as there is for C. The mean variation between adjacent pixels is called the roughness, by analogy to its visual interpretation by the microscopist. The slope of the brightness vs. distance plot (the correlation between brightness difference and scalar distance on the surface of the object) is a measure of the texture. These parameters are useful for comparison of uniformity within objects viewed in a transmission image, or of surfaces on different samples, or different regions of one sample (for instance, to compare the roughness of fracture surfaces near their origin and elsewhere).

Chapter 7

Special Techniques

In addition to the relatively standard techniques that have been discussed in the preceding chapters, there are a number of techniques that are special, in that they apply to (or, are applied to) images of non-ideal specimens, or to non-ideal images of specimens, or are simply still somewhat experimental. A few of these, which the routine user of stereological methods may have occasional need to employ, will be introduced here. Also, methods which are not strictly stereological but are nonetheless concerned with the determination of three-dimensional information from images are presented.

Preferred Orientation

The assumptions behind the relationships between measured and structural parameters in the preceding chapters began with the idea of a random sample of the structure, usually a plane section on which measurements are made. But many real specimens, whether biological or materials, are not random internally but have a degree of order that often manifests itself as an organization of structure that is different in different directions. For instance, muscle tissue, stems of trees or plants, metals that have been cooled in ingots or rolled into shapes, and geological formations, all look quite different when sectioned parallel or perpendicular to the principal orientation of the structural components (muscle fibers, layers of cells, elongated metal grains, or layers of particular rock composition). If measurements are made on any particular section plane, in ignorance or disregard of the true nature of the three-dimensional structure, the results can be extremely misleading about the size and shape of components in the material.

We will see later in this chapter that there are ways to comprehensively describe entire three-dimensional structures using serial sections. For our purposes here, it will be sufficient to imagine that we can perform measurements on more than one section plane (for instance, on three orthogonal planes), or in more than one direction in the material. Of course, it helps a great deal if the investigator suspects (from independent knowledge about the specimen) that there may be preferred orientation, and can make an intelligent guess at the likely orientation (for instance, growth directions, lines of force, action or deposition, or directions that correspond to external, macroscopic surfaces which may be related to internal structures). Comparisons between measurements made at different angles, or between lengths of features classed as a function of angle, can be made using the analysis of variance test described in Chapter 1 to determine whether a significant anisotropy is present, at least on the surface being measured.

121

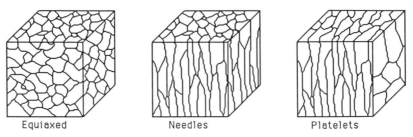

Equiaxed   Needles   Platelets

Figure 1. Examples of some of the kinds of preferred orientation.

Figure 1 illustrates a few of the possible internal cell or grain shapes that may be present. Some surfaces are the same in the drawings, but the other surfaces are quite different. The direction of elongation can usually be estimated from knowledge about the sample (for instance, the growth direction in a plant or the rolling or forging direction in a metal). If cut or polished sections can be prepared on all three surfaces, then measurements made on each surface will reveal the anisotropy because the cell or grain size will be different on each surface. Also, the aspect ratio of the features, and plots of the orientation of the length axis, can not only reveal the extent of the preferred orientation, but also can provide characterization so that different specimens may be compared.

Consider a section image in which the structure is characterized by parallel straight lines. If randomly oriented test lines are drawn on the image, and the number of points per unit line length P_L are counted, the value will vary as

$$P_L = 1 / d \cdot \sin{(\vartheta)}$$

where d is the mean line spacing in a direction perpendicular to their orientation.

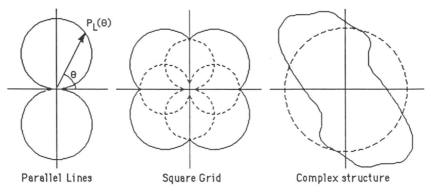

Parallel Lines   Square Grid   Complex structure

Figure 2. Rose patterns of the number of intersections per unit line length as a function of direction (Underwood 70), for a set of parallel lines and a square grid, and for a typical complex structure with preferred orientation.

Of course, it is unusual for structures to be completely or perfectly oriented. One possibility is that there are several lines of symmetry in the material (for instance, twinning along crystallographic axes in mineral or metal grains). In this case, the P_L patterns from the various sets of lines are superimposed. Figure 2 illustrates the resulting plots of P_L versus angle for the case of one set of parallel lines, or for two sets forming square patterns. The superposition of many different sets of symmetry axes makes the polar plot of P_L circular, while variation in the spacing of sets of lines in different directions produces plots of these "rose" patterns that demonstrate the degree and direction of the preferred orientation. For real structures, a comparison of the measured P_L values to the expected value for a randomly oriented structure (a perfectly circular rose pattern) can be straightforwardly made using the statistical methods shown in Chapter 1 (the statistical precision of the P_L value comes directly from counting considerations).

Without making the detailed measurements needed to construct an entire orientation rose, it is possible to learn much from measurements in two directions: parallel and perpendicular to the axis of preferred orientation. This is only practical because the human eye is generally very good at determining that direction. If a structure is made up of oriented (parallel) lines and random (isometric) lines superimposed, then we may take from the Buffon needle problem the relationship for the expected probability of intersecting the lines. The line length per unit area L_A for the random lines is $\pi/2\ P_{Lparallel}$ if we make the measurement P_L using test lines that are parallel to the orientation axis (and hence which never intersect them). If the test lines are oriented perpendicular to the orientation axis, the value of P_L will be greater, and the value of L_A for the oriented lines only can be obtained by subtracting $P_{Lperpendicular} - P_{Lparallel}$. Combining these relationships gives (Underwood 70)

$$L_A = P_{Lperpendicular} + 0.571 \cdot P_{Lparallel}$$

for the total line length per unit area. So by performing two counts of P_L, in two directions, the total line length per unit area can be determined. The degree of orientation is then given by the ratio of the oriented to the total value

$$\Omega_{12} = (P_{Lperp} - P_{Lpar}) / (P_{Lperp} + 0.571 \cdot P_{Lpar})$$

The subscript *12* indicates we are dealing with lines (1-dimensional) in planes (2-dimensional). Note that this value has the dimensions of a pure ratio. It is usually expressed as a percentage. Remember too, that while we have spoken of lines in a plane, these lines are normally the grain boundaries, cell walls, or other discontinuities in structure that we have been measuring all along.

To determine the degree of orientation of lines in space (eg. dislocations, microfibers), we apply a similar technique. Two perpendicular planes are examined, and the number of points where the 3-D lines intersect the planes are counted. Then the degree of orientation is

$$\Omega_{13} = (P_{Aperp} - P_{Apar}) / (P_{Aperp} + P_{Apar})$$

For surfaces in space, the value S_V is estimated by counting P_L (which has the same dimensionality) on sections that are parallel and perpendicular to the orientation axis. For systems that are "linearly oriented" (ie. arrays of parallel needle-like grains or muscle fibers with an equiaxed structure on a plane perpendicular to the orientation axis), the degree of orientation is given by

$$\Omega_{23lin} = (P_{Lperp} - P_{Lpar}) / (P_{Lperp} + 0.273\,P_{Lpar})$$

Arrays of pancake-like platelets or other generalized 3-D features give rise to a somewhat more complicated situation; there are three combinations of directions, and hence three planes on which to count P_L, and three Ω terms. They are all derived similarly to the cases above, and the coefficients all arise in the same way (the generalized Buffon needle problem in three dimensions).

$$\Omega_{23a} = (P_{Lperp} - P_{Ltrans}) / (P_{Lperp} + 0.429{\cdot}P_{Lpar} + 0.571{\cdot}P_{Ltrans})$$

$$\Omega_{23b} = (1.572 \cdot (P_{Ltrans} - P_{Lpar})) / (P_{Lperp} + 0.429{\cdot}P_{Lpar} + 0.571{\cdot}P_{Ltrans})$$

$$\Omega_{23c} = (P_{Lperp} - 1.571{\cdot}P_{Lpar} + 0.571{\cdot}P_{Ltrans}) /$$
$$(P_{Lperp} + 0.429{\cdot}P_{Lpar} + 0.571{\cdot}P_{Ltrans})$$

where the parallel, perpendicular and transverse directions define the plane orientations with respect to the principal structural orientation of the material. The "overall" anisotropy may be estimated as the square root of the sum of the squares of the three terms.

Note that these terms actually include less information than the rose, as they compare intercept lengths in only 2 or 3 directions. All of the measures of preferred orientation presented here may be useful to describe the degree of orientation, for comparative purposes, but they do not describe its nature. They also depend upon finding the "natural" coordinate system of the material with the polished faces, for instance parallel to a surface or aligned with a growth, rolling or drawing direction in a specimen.

Fractal dimensions

Very little has been said so far about the perimeter of features, from which we obtain, for instance, the surface area per unit volume. It was noted that the boundaries which define surfaces around embedded phases in a matrix may be broadened somewhat by etching, staining or other processes that increase image contrast, but we have ignored their thickness except in cases where, as discussed in chapter 4, they become a feature (a membrane or lamella) of sufficient thickness to be measured and described as part of the material volume. There is another problem with boundaries, however. It has to do with the fact that in nature, most boundaries are not planes or simple smooth curves (our by-now-familiar ellipsoids), but are instead rough and irregular. Furthermore, the amount of roughness and irregularity that we can see is generally limited by our image resolution, and if we could increase the magnification, the amount of perimeter we would see in the image would increase.

The simple models that have been used so far for shapes, such as spheres, ellipsoids, and polyhedra (and their corresponding two-dimensional profiles) are useful for many purposes, including estimates of volume or size distribution, but certainly fall short in dealing with surface area, which is likely to be larger (perhaps much larger). One method for describing the irregular shape of feature profiles has been to "unroll" that shape by plotting distance from the centroid as a function of angle, and then to perform a Fourier analysis on the resulting curve (Beddow 77). Besides being computationally demanding, this approach has difficulty in dealing with shapes so irregular that the radius line may intersect the outline more than once.

A new approach to this problem, using what are known as fractal dimensions, requires us to think about non-Euclidean lines and surfaces. Euclid defined a "line" as a geodesic, that is the shortest path between two points. In our common three-dimensional space, this is the familiar straight line (although there are various non-Euclidean geometries, mostly based on other behaviors of parallel lines, where they may be other smooth curves). For our purposes, we will consider lines to be instead a boundary between two regions, which has no internal tension and does not strive to be straight. The constraint that the line be "smooth" (that is, have derivatives at every point) is specifically relaxed, so that the line can be "crinkled."

The extension of this approach to surfaces is straightforward. Instead of flat or smoothly continuous "Euclidean" surfaces, we will be particularly interested in things that can be crinkled up in ways that may better describe many "real" surfaces. (Of course, there are some natural surfaces that really are smooth and Euclidean, particularly facets that develop because of the crystallographic lattice of materials, or membranes that assume a least-energy configuration in response to surface tension.)

The particular type of crinkled-up behavior that is of most interest to us here has a peculiar form called self-similarity. This simply means that at any magnification at which we choose to view the line or surface, it looks the same. Whatever measurements we can make to describe the roughness and its scale will be independent of the scale. This imposes a significant constraint on shapes of features. Recall that the formfactor described before ($4\pi\, Area\, /\, Perimeter^2$) can be the same for two very different shapes, such as a daisy and a fuzzy disk (see Figure 3). Fourier analysis of the frequency components of these shapes (sometimes used in image analysis) would show that the disk outline contains primarily high frequencies (short wavelength), and the daisy primarily low ones. Self-similar outlines contain all frequencies equally. Interestingly, these kinds of shapes are actually quite common in nature.

Figure 3. Two shapes with similar form factors but very different frequency components.

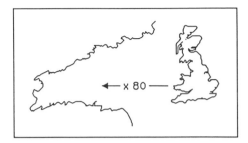

Figure 4. The west coastline of Britain at two magnifications, one at a scale 80 times larger than the other.

This effect was first noted in two dimensions, by a somewhat eccentric British mathematician named Richardson. He measured the borders of several countries and landmasses, but the best known example is the west coast of Britain (Scotland, England and Wales) as shown in Figure 4.

If the length of the boundary (the coastline) was determined by swinging along a map of the country with dividers set at some arbitrary distance, say 1 kilometer, a total length was obtained. Then if the divider distance, or stride length, was reduced, and the same operation repeated, the measured length was greater because the measurement was able to follow more of the irregularities of the coastline. Repeating the operation with finer and finer steps (and maps of appropriate scale) would cause the length to continuously increase, so that in effect, the length of the coastline would be expected to become infinite at a fine enough scale. Furthermore, Richardson noted that over a considerable range of stride lengths, and for a variety of borders, both natural and man-made, the slope of the plot of measured length versus stride length was constant (see Figure 5) on a log scale.

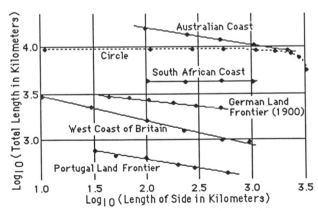

Figure 5. Richardson's plots (61) of the length of various geographical boundaries versus the distance used for measurement (the length of the side of the polygon used to fit the boundaries).

Figure 6. A Koch island with a perimeter whose fractal dimension is 1.5

An example of the effect of increasing perimeter without increasing area is shown in Figure 6. This type of feature is called a Koch Island, after the mathematician who proposed them (Von Koch 04).

All the features are drawn with the same area; you are asked to imagine that in each of the different images you have improved your image resolution to be able to define smaller details. The perimeter on each successive feature is increased because of the patterned irregularity that is introduced on each straight line segment of the boundary. In fact, the amount of perimeter increases by a factor of 1.5 from each feature to the next, and there is no limit (at least down to atomic dimensions) to how far we might extend this process, so the length of the perimeter is undefined. The mathematicians who considered shapes like these were horrified at the behavior of such lines and shapes because without smooth, continuous derivatives, they could not be easily handled by conventional techniques.

Mandelbrot (83) has developed these ideas further into a useful concept. If the increase in measured length with improvement in measuring resolution is uniform (a straight line on a log plot, as shown in Richardson's data), the feature is said to be self-similar. He has shown that many natural objects, including the surfaces of particles as viewed in the SEM, have this character at least over a substantial range of distances. The consequence is that it is not really possible to state what the amount of boundary or surface area is. For the Koch island shown in the figure, the fact that the perimeter increases by a factor of 1.5 for each halving of the measurement distance has led Mandelbrot to describe this particular boundary line as having a dimension not of 1 (a straight line) or 2 (a plane) but 1.5. In other words, the line has a fractional dimension which reflects its ability to fill a plane, and hence it is called a "fractal."

Depending on the nature of the irregularity which we might introduce on the Koch island, the rate at which the perimeter increased with each step could vary. Figure 7 shows several such curves. In each case, the area of the feature would stay

| 1.25 | 1.393 | 1.5 | 1.661 |

Figure 7. Irregularities that can be substituted on the Koch island, with different fractal dimensions.

the same as each straight line segment in the feature boundary was replaced by the new curve, and then the sequence repeated at a finer scale, and so on, but the rate at which the perimeter would increase is given by the values shown. These values are related to the slopes of the Richardson plot that one would obtain for these features. Higher fractal dimensions reflect the fact that as the substitutions of finer scale irregularities are made, the line spreads out faster over the plane. A fractal dimension approaching 2.0 would cover the entire plane, while one close to 1.0 would remain nearly a line.

Of course, in real features the irregularity is not so regular. It is possible to apply a random pattern to increase the perimeter along edges, with a mean fractal dimension, and Mandelbrot has shown some examples of this technique that produce features with stunning realism (Lucasfilm has extensively developed and utilized these methods, to produce realistic but un-earthly landscapes for movies). The concept of self-similarity simply means that on the average, the increase in boundary length is uniform as resolution is increased (for instance by working at progressively higher magnifications).

Generating a fractal boundary that is "random" can be accomplished rather straightforwardly by using the first type of generator shown in Figure 7, while varying the distance by which the new points on the line are offset. The outlines shown in Figure 8 began with polygons formed by generating a random series of points on a circle. Each side was then divided into 2 parts and the midpoint displaced either in or out by a random distance, proportional to the length of the side. This process was repeated 6 times, so that the side ultimately became a series of 128 short segments. The total figure has a fractal dimension that is easily determined by constructing a Richardson plot (obtained by summing the length of the perimeter for each of the halving steps). The mean displacement of the midpoints can be varied from zero up to half the length of the line segment whose midpoint is being displaced, to produce fractal dimensions from 1.0 (a Euclidean polygon) to about 1.25, and the shapes created are very similar in appearance to those of some real objects (cornflakes or dust particles or islands, for example).

For some applications, it is satisfactory to visually compare the "roughness" of real objects (typically viewed in the microscope) with generated outlines having known fractal dimensions, to determine an approximate value. This may be adequate for comparative purposes, but measurement of the dimension using a Richardson-type plot is desirable since it also provides confirmation of the self-similarity of the object's outline, established by the linearity of the plot.

To apply this method to real structures, such as particulates whose surface area may be important (as a substrate for chemical reactions or a site for chemical diffusion, for instance), measurements are made using a series of stride lengths and the total dimension is plotted. Usually, the perimeter thus measured is not a perfect integral multiple of the stride length, and some partial length is left over at the end. The resulting real number (integral number of strides plus a fraction) times the length is then the perimeter. If this is plotted against the stride length (λ), we obtain a Richardson plot as shown in Figure 9. The plot deviates from ideal behavior at short stride lengths because of the finite resolution of the image being measured, and at large stride lengths which become a significant fraction of the size of the profile.

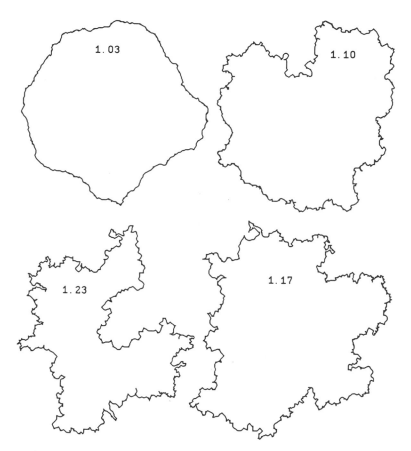

Figure 8. Random fractal outlines generated as described in the text, with varying fractal dimensions.

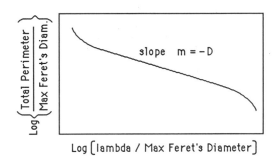

Figure 9. Schematic Richardson plot for a particle outline.

To normalize the data, it is customary to express both measurements as ratios to a characteristic length for the feature, usually the maximum Feret's diameter. This is not necessary to determine the fractal dimension itself, but is useful to compare plots of features of different absolute size. It is also necessary to adopt a convention for dealing with cases along the periphery when swinging the dividers from one point can intersect several points along the boundary. For instance, we might decide in that case to use the point farthest from the original, and continue the process from there.

Once the log plot has been obtained, the fractal dimension of the boundary is obtained from m, the slope of the linear portion of the plot as

$$dimension = 1 - m$$

Because we shall also be applying the concept of fractal dimensions (and Richardson plots) to surfaces, which have a dimension between 2 and 3, it is most useful to deal with the fractional part separately from the integer. For a similar plot of log total surface area (vertical scale) vs. the log of the area element used to measure it, the slope will still be $-m$, but the dimension will be $2 - m$. This is handled most simply by writing the dimension as $1.D$ or $2.D$ where the value of D ($= -m$) is the decimal fraction, between 0 (the Euclidean limit) and 0.999... The upper limit cannot actually exceed 1; for the Koch islands shown above, if the generating element had an increase in length so great that the fractal dimension was 2.0, then the resulting shape would spread without limit across the plane. Also, the curve would have to be self-intersecting, and the boundary would no longer unequivocally separate the inside (feature) from outside (surroundings).

Comparison of the roughness of different classes of features can then be made using this parameter. The value of m or D from the Richardson plot is obtained by performing a linear least squares fit to the data points, and the value and its standard deviation may be compared to values from other specimens using, for instance, the analysis of variance (ANOVA) method from Chapter 1. This provides a quantitative method for comparison of different shapes.

The measurement of perimeter by hand, as described, is very tedious. Computer-assisted determination of the fractal dimension of an outline may be carried out in several ways. The most straightforward is available when the feature outlines are available as a series of point coordinates around the perimeter, as would normally be obtained if the feature were outlined using a graphics tablet digitizer to delineate the boundary manually. This outline is a many-sided polygon, and with most types of equipment, the spacing of the points is fairly uniform (depending perhaps on the speed of motion of the operator's hand in moving the stylus around the outline, or on a sensitivity setting in the software). If is easy to sum the perimeter of the polygon using all of the points, and then to repeat this with every second point, and then every third, and so forth (Schwarz & Exner, 80). When this approach is used, it is best to average the perimeter determined with each point spacing using all possible starting positions, to minimize bias due to sharp fissures or projections on the outline.

The perimeter will in general decrease as the polygon is made coarser (in the Euclidean limit it would not change). Plotting the log of perimeter vs. the log of the side length of the polygon (which is adequately approximated by the perimeter divided by the number of points used) produces a Richardson plot, from which the fractal dimension D can be determined. This procedure obviously cannot be continued until there are too few points, because the area of the feature begins to change. Note that the plot proceeds from the finest resolution available, namely that of the original outline, upwards to a limit imposed by the feature size. It is not possible to proceed to finer dimensions for the stride length, because there is simply no information available at a smaller scale than the original point spacing.

It is possible in principle to perform the same operation with raster-scanned images, since we have seen before that the pixels which touch and form the feature can be converted to a chord table, and ultimately to a list of points which can be sorted (with some difficulty) into a sequence of vertices that define the boundary polygon. This requires considerable computation, however, and a simpler method is much faster to use. Among the parameters for each feature normally determined from the original pixel image, the perimeter is computed by summing the distance (using the Pythagorean theorem) between ends of adjacent chords. If this perimeter is repeatedly summed as the pixel image is repetitively coarsened, a Richardson plot can again be constructed.

The coarsening is begun by replacing each block of 4 pixels (2 on a side) by a solid block in which the pixels are either all on or off. If the original block has 0 or 1 pixels on, it is set entirely off, and conversely if 3 or 4 pixels are on, all are set on. For the case where two pixels are set and two are not, there are 6 possible configurations. Three of these are arbitrarily chosen to produce blocks that are entirely on, and the other three result in all pixels being turned off. The result is an image with a more blocky appearance, and a shorter perimeter (but, on the average, the same area). Figure 10 shows an example of this process.

The process can then be repeated. The actual routines are much the same as those used for etching and plating. Either a series of 3x3, 4x4, 5x5, etc. blocks can be used from the original image, or for greatest simplicity, the same rules just used for the 2x2 block can be reapplied using 4x4 blocks in which the original 2x2's are treated as single pixels. Repeating this 4 or 5 times gives enough points for a good determination of D. As for the polygonal outline, this only proceeds from the finest

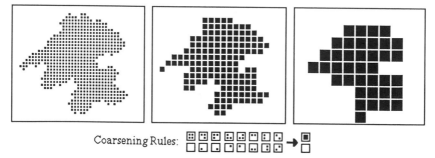

Figure 10. A feature made up of pixels, and the result after it is coarsened twice using the rules shown. The reduction in perimeter after several repetitions can be used to determine the fractal dimension of the outline.

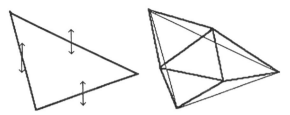

Figure 11. Generator for a random fractal surface. The midpoints of the sides of the triangle are displaced vertically in proportion to their length, and produce vertices for new, smaller triangles.

resolution of the original image to larger dimensions, as there is no information available at a finer scale. The stride length is usually taken as the edge dimension of the block, starting with one pixel and proceeding to multiples thereof. This is slightly biased for the usual case of a square grid of pixels, since some of them are diagonally adjacent, and hence are spaced 1.4 times as far apart, but this merely shifts the plot sideways and does not change the slope. A hexagonal array would be preferred to keep the neighbor distance uniform, just as in the plating and etching situation, but is rarely used because raster scan hardware is much simpler when it uses square pixels.

Other methods have also been described for determining fractal dimensions for outlines in rasterized two-dimensional images. Generally, they all rely on some type of plating process. For instance, in principle it is possible to pass a circle (hard to approximate with square pixels) along a surface and determine the area of the band that is swept out (Coster & Chermant, 85). The variation of this area with the circle diameter (which is effectively the stride length) then gives the fractal dimension. The "coarsening" method described has been shown to yield results which agree closely with manual methods, or with the known fractal dimensions of outlines generated as shown above.

The same principles apply to surfaces as to outlines. Generated random fractal surfaces can be easily produced by beginning with a triangle, and then displacing the midpoints of each side up or down by a random distance. This creates four smaller triangles replacing the original, as shown in Figure 11, and they have a greater surface area. These triangles can then be further subdivided to produce a model of a rough surface. Varying the amount of displacement controls the rate of increase in surface area as the triangles become smaller, exactly analogous to the outlines shown before. The result is a series of surface profiles with different fractal dimensions, as shown in Figure 12.

Measuring the area of a surface is much more difficult than is measuring the length of an outline. To determine the fractal dimension of a surface, it is necessary to have surface areas determined using a range of measuring elements of different sizes. It is possible to do this with triangles, using the same method as will be described for plotting surface contour maps from a series of points whose elevations are determined stereoscopically. However, this is extremely tedious even with computer assistance. There are also a few cases in which altogether different methods can be used, such as BET measurements of surface area using different molecular species (with varying sizes). However, for the general case, no direct method for measuring surface areas is really suitable.

Consider, though, the stereological principle introduced in Chapter 3: It is possible to perform measurements on two-dimensional intersections through solids from which information about the three-dimensional shapes can be obtained. For instance, the surface area per unit volume S_V may be determined from the length of

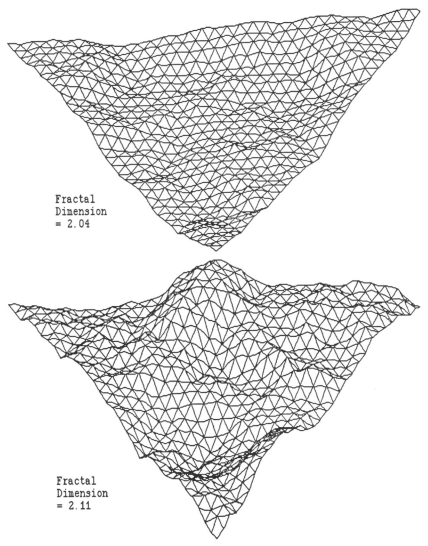

Fractal
Dimension
= 2.04

Fractal
Dimension
= 2.11

Figure 12. Surfaces generated using the method of Figure 11, with varying amounts of displacement and hence varying fractal dimensions.

Figure 13. A generated fractal surface (dimension 2.18) intersected by a plane. The intersection contour has a fractal dimension of 1.18.

boundaries per unit area B_A, or even the number of intersection points per length of line P_L. This encourages us to try to relate the fractal dimension of a surface (a value between 2 and 3) to that of its intersection with a plane (on outline with a fractal dimension between 1 and 2). One might consider as a familiar case the shoreline around an island. Islands which are generally rather rough and hilly would be expected to have a shoreline that is more irregular than islands which have a more gradual topography. Fractal dimensions allow us to quantify these descriptive terms.

For the ideal case, Mandelbrot has shown that for surfaces generated using randomized self-similar fractals with a dimension of the form $2.D$ ($0 <= D <= 1$), plane sections through the surface produce outlines with a fractal dimension $1.D$. Figure 13 shows an example in which a random fractal (generated as described in Figure 11) is intersected with a plane ("sea level"), producing an outline ("shore line") whose fractal dimension can be measured. This exciting (and not particularly obvious) result means that measurements on planar sections using the methods already described can be applied directly to describe surfaces. Note that this applies only to sections through three-dimensional features, and not to projected or shadow images as are often viewed in the microscope. For that case, the outline will have a value of D that sets a lower limit to the value for the surface.

Dimensional analysis of the relationships between surface area, volume and length are somewhat different with these non-Euclidean boundaries than those with which we are generally familiar. For instance, the "form factor" described earlier for the shape of features was taken as $4\pi \, Area \, / \, Perimeter^2$, which is dimensionless. In general, we would expect any ratio of $volume^{1/3}$, $area^{1/2}$, and $length$ to be dimensionless.

For Euclidean shapes, this might take the form of expecting $volume^{1/3}$ to be proportional to $surface \, area^{1/2}$ for a series of features. However, very different relationships are actually observed in the real world. For example, for mammalian brains, the actual relationship is that *volume$^{1/3}$* is proportional to *surface area$^{1/2.76}$*. This is another way of saying that the surface of the brain has a fractal dimension of 2.76, which would be revealed by section outlines with a fractal dimension of 1.76 (in the Richardson plot of either log total surface area vs. log of area element, or log total outline perimeter vs. log stride length, the slope is -0.76, and $D = 0.76$).

Many other similar relationships have been quantified for living things (McMahon & Bonner, 83), usually by making log-log plots of parameters with very different dimensions and units, and finding straight line relationships. Some do not fit the simple slopes expected from Euclidean geometry, which in this circumstance is described as the ideal allometric behavior. The deviant cases may involve fractal behavior for such things as bending behavior of trees (which have fractal branching patterns) and speed vs. size for swimming fish (the boundary layer, which contributes most of the resistance, may have a fractal nature).

Examples from other fields are also available. Consider, for instance, the relationship between the length of a river system (which will depend on the length of the yardstick used to measure it), and the area of the drainage basin (which has conventional units of area, and is not sensitive to the scale of measurement). It has been shown that the square root of the basin area is proportional to the straight line distance from source to mouth, but that the length of the river is proportional to the area raised to the power 0.6 (rather than the expected 0.5). This is equivalent to stating that the fractal dimension of typical rivers is 1.2. Notice that this area vs. length relationship can be used to determine a fractal dimension without the need to perform repeated measurements on the same river with different yardsticks. The similar use of area vs. volume relationships for particles viewed in the microscope is often a more efficient way to determine the fractal dimension of the surface than direct measurement of section profiles.

As a practical example of alternate ways to determine the fractal dimension of objects, Figure 14 shows a plot of the measured mass (weight) of clusters of aggregate particles versus the maximum dimension of the aggregate. Both axes are normalized to an individual particle (which means that the vertical axis is actually the

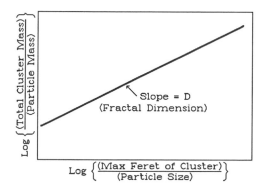

Figure 14. Plot of number of identical particles in an aggregate versus its maximum dimension has a slope equal to the fractal dimension of the aggregate surface.

number of particles in the aggregate), but this does not change the overall interpretation (which is appropriate even for clustering of individual atoms). Depending on the conditions under which the aggregates form, and possible constraints on the way that individual particles can migrate toward or attach to the aggregate, the surface ruggedness can vary widely (consider branching, dendritic structures and perfectly faceted crystals as extremes). In this log-log plot, the slope gives the fractal dimension of the surface of the aggregate directly, and it is just equal to 2 for cases in which ideal Euclidean objects (such as perfect crystals) form.

Mandelbrot (83) has cited examples of fractal behavior in the condensation of droplets of liquid from vapor, the area/perimeter relationship for rain clouds, and for membranes in the lung (fractal dimension 2.17), intracellular membranes around mitochondria (fractal dimension 2.09 for the outer membrane and 2.53 for the inner membrane), and for the structure of endoplasmic reticulum (dimension 1.72 approximated as a fractal line).

Fewer examples seem to be available yet for materials science, perhaps because non-fractal behavior is more common there (due to crystal structure, surface tension, and other thermodynamic constraints on boundaries and surfaces). However, a correlation between the fractal dimension of outlines (obtained by plating and sectioning brittle fracture surfaces in glass ceramics and alumina) with the measured fracture toughness of the materials (Mecholsky & Passoja 85), and numerous attempts to model fracture propagation and the growth of deposited aggregates using random fractals (Laibowitz et. al. 85) are dynamic indications of studies now being performed.

Kaye (84) has pointed out that ideal fractal behavior (self-similarity as evidenced by a linear Richardson plot) only occurs over a finite range of distance, and in fact that there may be different slopes in different size ranges. These may correspond to real structures in materials, ranging from the scale of the crystalline structure to the size of grains, pore structures, or macroscopic machining marks in components. Nevertheless, correlation of fractal dimensions of fracture surfaces with mechanical properties, or of particle surfaces with adhesion, etc., may offer valuable insights into materials characterization and behavior. Work with fractals is quite new, and more development in the theory and interpretation may be anticipated.

Serial sections

The stereological and stereometric methods described so far deal for the most part with observations and measurements made with lines and points on random planes through specimens. There are some details of three-dimensional structure and organization that cannot be deduced from such measurements, including 3-D shape and the relationship between various associated structures. One technique that can be applied in some cases is that of serial sectioning. This has proved especially useful for biological specimens, not only because the structures there are especially intricate, but also because the specimen preparation techniques are appropriate.

In serial sectioning, as shown schematically in Figure 15, the specimen (perhaps embedded in a matrix that is easily cut and relatively transparent to light or electrons) is cut into a series of parallel slices, each of which can be examined and photographed in the light or electron microscope to show the internal structure,

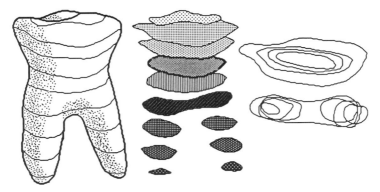

Figure 15. Schematic diagram of serial sections of an object, with normal views of the superposition of a few of the section outlines.

features and their shapes. In some cases with transparent samples in the light microscope, the same effect as serial sectioning can be obtained utilizing the shallow depth of field of light optics at high magnification. A series of images are recorded as the plane of focus is shifted through the sample in the depth direction.

While each photograph shows the usual details, it is only by combining the photos that the 3-D structure is shown fully. This may be done by printing the photos onto transparent overlays, or by using them to make solid models of the features by cutting out styrofoam, or other materials, to the indicated shapes. It is also becoming common to digitize the outlines of features (usually with human delineation of the features, not only because biological images are so complex, but also to select just those features and boundaries that are of interest). This has an advantage over physical models in permitting internal structures to be viewed, or even to permit viewing from the inside.

Then, whether the outlines are stored in computer memory, or exist as transparencies, it is necessary to align them in proper registration. This may not be an easy task if the orientation and alignment of the sections has not been carefully maintained (it becomes especially difficult when rotation between sections may have occurred). Once the outlines are lined up, the relationships between structures become evident. Figure 16 shows a very simple diagram of a few serial sections, in which the position of the nucleus and several mitochondria inside a cell are shown. In individual planar sections, mitochondria almost always appear as approximately elliptical features. Serial sections demonstrate that they are not shaped like footballs, but are long, convoluted structures.

Computer reconstruction of the images from the stored outlines is especially powerful because it is possible to rotate the structure and view it from any point in space, even locations within the structure which would be inaccessible with a solid model. It is also possible to perform integration of volumes and surface areas for individual features, from the section images (the volume is simply the sum of areas times thicknesses, and the surface is simply the sum of perimeters times thicknesses;

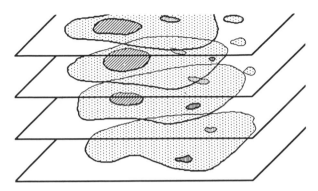

Figure 16. Illustrative example of serial section information, showing internal features.

Simpson's rule should be used to perform the integration). However, most of the use of serial sections is not for quantitative measurement, but rather to allow study of complex 3-D relationships between internal structures. Serial sections may also be used for systematic sampling of structures, as opposed to random sampling. This may be preferred, or even necessary, for three-dimensional structures such as biological organs which are not randomly arranged internally.

Topology

One of the kinds of information that serial sectioning can provide, which is normally not discernible on single sections, is the detailed shape of "things" (features, pores, particles, tubules, etc.) in the specimen. The description of shape involves the field of topology. This makes possible comparisons of the connectivity of pores in ceramics, blood vessels or other ductwork in kidney, roots of plants, neurons in the nervous system, and other complex structures, as well as the shape of grains in metals.

For instance, if we use a polyhedron as a model of grain shape, then for C = the number of corners, E = the number of edges, F = the number of faces, and G = the number of grains, these quantities are linked by the relationship

$$+ C - E + F - B = 1$$

If we assume (for the case of no voids between grains, which would themselves be a different class of polyhedra) that each corner is shared by 4 grains, each edge by 3, and each face by 2, then we can calculate the number of edges around each face as:

$$n_{mean} = 1 / F S_n F_n$$

where F_n is the number of faces with n edges. For a specified shape, this value can be calculated. For a 14-sided tetrakaidecahedron (which is one of the space filling shapes introduced in Chapter 4), the value for n_{mean} is 5.143, meaning that in general we should see grain outlines with this number of edges. Measurements on a variety of real structures (bubbles, metal grains, cells in plant tissue, etc.) show distributions of n from 3 to 8, with a mean typically just greater than 5. This is an encouraging confirmation of our models for grain shapes.

In metals, for example, real grains may have an average number of faces of about 14, but are clearly not uniform in size or shape. Surface tension (related to curvature as mentioned in Chapter 3) requires that at equilibrium, the three edges of adjacent grains meet at an angle of 120 degrees, and at corners where four grains meet, the tetrahedral angle should be 109.5 degrees. Real structures, while not in perfect equilibrium, generally show angles that cluster closely around these ideals. Surface tension considerations also require that polyhedra (grains) with more than 14 faces have grain boundaries that bow outwards (are concave), which provides a driving force for grain growth, while grains with fewer than 14 faces have the opposite curvature. Hence, it is expected that in grain growth, grains with more than 14 faces should grow at the expense of ones with fewer faces. The curvature of the boundaries should also depend on the number of faces. These effects have been observed in metals.

Topology is particularly concerned with descriptions of the shape of networks of lines and surfaces in space. These descriptions ignore what we may conventionally think of as "shape" and instead reduce the network to its simplest form of lines (or surfaces) and connections. Thus a cube and a sphere are the same, from a topological point of view, because one can be stretched into the other without the need to cut anything, but a sphere and a doughnut are distinct because one cannot be stretched into the other. The degree of connectivity of a feature is given by its "genus," the number of cuts one could make in the reduced network without separating the feature into more than the original number of parts. For instance, in Figure 17, the original feature is first shrunk down to its essential connections, and then some of these are cut. For the example shown, up to three cuts can be made without disconnecting any of the nodes (places where the linear segments connect). Hence, this feature is of genus three.

The connectivity of features is important in many disciplines. For instance, the three observed stages of sintering of ceramics, which have different kinetic behavior, can be linked to different degrees of connectivity in the pore structure

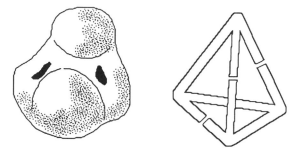

Figure 17. A multiply-connected 3-D feature and its reduced form, with three cuts.

between the particles. Likewise, the number of branches and nodes in a feature's skeleton often correlate with various physical properties.

Similar descriptions can be given for bodies with one, two or three dimensions in space. One-dimensional (lineal) features, often called tubules, and two-dimensional surfaces (sheets or "muralia") can be investigated using the measure of curvature K_V introduced in Chapter 3 (and determined from a net tangent count on a section plane). For particles, $K_V = 2\pi N_V D_{mean}$ where D_{mean} is the mean tangent diameter as defined before. for tubules, $K_V = \pi L_V$ where L_V is the length of tubules per unit volume. For muralia, $K_V = \pi/2\, L_V$ where L_V is the length of edges of the sheets (the total perimeter), and $L_P = (\pi/2)^2\, S_V\, /\, K_V$ gives the lineal mean intercept in the surface of the sheet. Dehoff (78) has discussed other topological parameters that can be obtained from the curvature.

Stereoscopy

Another three-dimensional method, which often produces more conventional quantitative measurement data, is stereoscopy. The principle behind this technique is familiar to most people already, since they use it constantly in daily life. We view most relatively nearby objects with two eyes, which must rotate in slightly to center the object on the fovea. The angle by which the eyes are rotated in sends cues to the brain that are interpreted to estimate the distance to the object. Many demonstration experiments such as trying to touch your fingertips together with one eye closed demonstrate the effectiveness of this simple trick of binocular vision.

In the early part of this century, photographers discovered the technique, and built special cameras to record two photographs of the same scene, spaced about the same distance apart as a person's eyes. The result was a pair of photographs (or, more commonly slides) that, when viewed in a "stereopticon," recreated in the eyes and mind of the viewer the original scene with all of the depth information.

The same method is commonly applied in the electron microscope (it is less appropriate to the light microscope, because the depth of field of light optics is usually to small to focus images of samples that have enough relief in the third dimension to be interesting). It is also used, with some trivial differences in the way the photos are taken, in aerial photographic mapping. In the TEM or SEM, two sequential micrographs are taken with the specimen tilted between them. The result is two pictures in which the relative positions of nearer and farther objects are shifted with respect to each other. Figure 18 shows the appearance of objects in the two views taken at a different angle. With an appropriate viewer, it is possible to see this parallax as vertical relief, and to measure it.

Left   Right

Figure 18. Stereoscopic views of a polygonal feature.

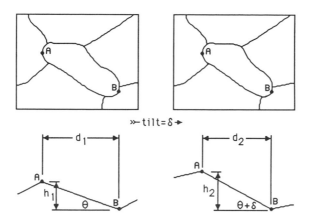

Figure 19. Stereoscopy measures the parallax between two photos to determine elevations.

Figure 19 shows the principle. Consider two points on a surface viewed in the SEM (the analysis is identical for two points within a thin section viewed in the TEM). In one picture, they appear with a separation in the horizontal direction of d_1. In the second one, after the sample has been tilted in that direction by a known angle δ, the points have moved (because one is closer to the viewpoint) and the new distance is d_2. Simple trigonometry then can be used to solve for the distance h_1, which is the vertical relief between them in the original view, or for Θ, the original angle of the surface which the points define. A thorough derivation and review is given in Boyde (73).

$$\Theta = \tan^{-1} [(\cos\delta - d_2/d_1) / \sin\delta]$$

$$h_1 = (d_1 \cos\delta - d_2) / \sin\delta$$

Notice that the angle is independent of the magnification, since d_1 and d_2 enter only as a ratio. The angle difference between the two images must be great enough to show some measurable parallax, but not so great that the eye cannot recognize common features in the two images. For very flat surfaces, angles up to 15 or 20 degrees may be used, but 5-10 degrees is more common.

In the case of aerial photos, or some SEM micrographs taken at rather low magnification, the parallax is generated by shifting the specimen horizontally rather than tilting it (or, by flying the airplane ahead at the same altitude). The calculations are similar, requiring instead of the angle difference between the two pictures, the distance from final lens to sample (or the plane's altitude) and the distance between the two photos (equivalent to the spacing between your eyes in normal stereoscopic vision). Referring to Figure 20, the elevation difference h between two points in given in terms of the parallax P by

$$h = WD \cdot P / S \qquad \text{where } P = d_1 - d_2$$

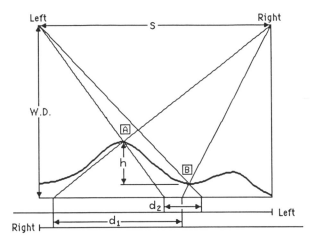

Figure 20. Schematic diagram of stereoscopic measurement of elevation h between two points *A* and *B*, using left and right images obtained by shifting the surface by a distance *S* at a working distance *WD*.

It may additionally be necessary, because of foreshortening of lateral distances in the image when large amounts of relief are present, to correct the X and Y coordinates of points in the image. This is done by calculating

$$X' = X \cdot (WD - h) / WD \text{ and } Y' = Y \cdot (WD - h) / WD$$

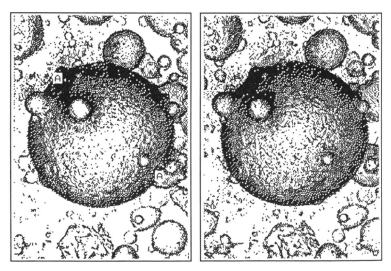

Figure 21. Stereo pair photographs of spheroidal nickel alloy powder. Taken with SEM, tilt 10 degrees. (Roberts & Page 81).

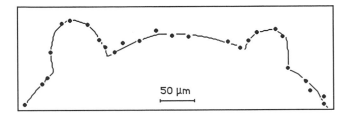

Figure 22. Elevation profile (*A–A*) across the powder particle in Figure 21

The manual measurement of the parallax is straightforward, but for many purposes it is desirable to automate this using a small computer and digitizing pad, as described before in the context of image measurement. If a series of points are measured, it is possible to form a traverse line across a portion of the sample or surface to record the elevation profile. An example is shown in Figures 21 and 22, in which points along line *A–A* were measured in this way to obtain the profile shown.

If repeated profiles are measured along a series of parallel lines, the profiles may be plotted as shown in Figure 23. This type of view, in which the individual profiles are displaced slightly to produce a quasi three-dimensional effect in the final image, is called an isometric drawing. It is especially useful for features with relief that runs primarily in one direction, as is the case for the semiconductor trace shown.

If enough points are measured, the entire surface can be mapped in elevation. Figure 24 shows an example of this, for a very simple specimen (a hardness indentation in a metal). The measured points are each recorded with the X,Y position, and the calculated height from the parallax measurements between the two pictures (in the figure, the original photographs have been reduced to stylized outline drawings of the most visible features to improve the clarity). The measured points are organized into triangles of neighbors, and within each neighbor the contour lines are drawn by linear interpolation. With enough points (typically 1000 or more), smooth lines are drawn which reflect the elevation contours of the surface.

Obviously, performing the measurement and drawing operations is most easily handled with a small computer. In this case it is only required to mount the two stereo pair photographs side by side, on a digitizing tablet, and then alternately locate with the stylus the same point on the left and right photos. If is rather easy (if somewhat tedious) for a human observer to do this, as the surface details on the pictures are complex and readily recognized. However, attempts to automate this

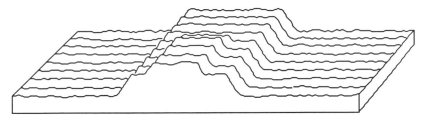

Figure 23. A series of elevation profiles across a trace on a semiconductor.

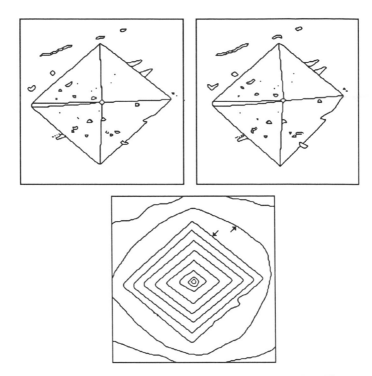

Figure 24. Contour map of metal surface with a hardness indentation. The contour lines have a 5 micrometer spacing, and the diagonal width of the diamond indentation is about 100 micrometers. Contour maps do not show the direction of slopes; the lines marked with arrows have the same elevation, indicating a raised crest around the indentation beyond which the surface drops away.

operation have met with only limited success, because the identical points on the two images do not have identical contrast (this is especially true when tilting specimens in the SEM, as the contrast changes both locally and with respect to the rest of the sample when the angle of inclination is changed).

An alternate method for determining surface relief utilizes the shallow depth of field of a light microscope. The specimen is mounted on the microscope stage, and as it is moved along a line, a human (or in some cases a microprocessor controlled focussing device) brings a series of points into focus, records the vertical motion required to do this, and from the series of points constructs and elevation profile of the surface.

Image averaging and autocorrelation

In some cases, the fine details in an image are not visually obvious. If the features are broken up, and masked by other random lines or objects, they may not even be discernible. But if there is a characteristic spacing at which similar objects repeat, it can often be found by a technique known as autocorrelation. If the image has been stored in a computer, there is a brightness value associated with each pixel. Autocorrelation produces a spectrum in which each point holds the sum of the products of pixel brightness values with other pixels in the image, and the horizontal axis of the plot is the distance between the rows of pixels that are multiplied. When random variation in the brightness patterns or spacings of features are present, the spectrum is featureless. But any repeating pattern will cause bright or dark regions to come into alignment at certain shift distances, resulting in peaks or valleys in the spectrum. Figure 25 shows an example of an autocorrelation spectrum. Note that the peaks show up at multiples of the distance between features.

This technique requires a fair amount of computing, and is only practical in the X and Y directions. More detailed information can be obtained from a Fourier analysis of the spectrum, which will also show spacings of repetitive features. Some images can take advantage of repetitive patterns in the image. Particularly for specimens of low inherent contrast, like many biological samples, when examined at or near the limit of magnification in the TEM, the noise in the image (due to the low number of electrons that form each point in the image) may mask detail. But when the structures, such as microfibers in cells, or atomic scattering contrast in material monolayers, appear in regular arrays, it is possible to combine the information from many repeating images by adding together the stored brightness values. This requires the ability to pick out portions of the image and translate them in X and/or Y, and perhaps also to rotate them. In fact, for features that have rotational symmetry, rotation and addition of a single feature image with itself can provide the same kind of improvement in image signal-to-noise ratio, and consequently in the ability to see fine detail.

Figure 26 shows an example of this technique, which is properly called correlation averaging (but sometimes referred to under the catch-all phrase of image processing). The image is a TEM micrograph of a patch of crystallinity in chlorinated copper pthalocyanine. Combining the multiple individual images, taking advantage of the mirror symmetry present, and contouring the resulting composite image, shows detail that is quite invisible in any single image (Saxton & Koch 82).

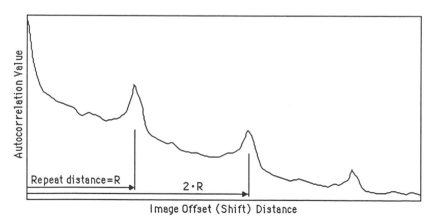

Figure 25. Example of an autocorrelation spectrum.

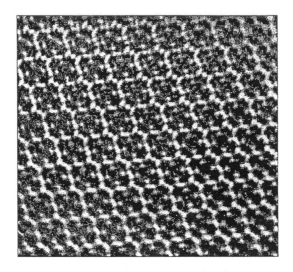

Figure 26. Top: a portion of a periodic image; Bottom: composite image after correlation averaging and contouring.

Autocorrelation is not necessarily restricted to use with grey scale images. It can be very efficiently applied to binary images, by shifting the image in X and/or Y directions and then *AND*ing with the original. The number of pixels which survive this operation, divided by the number of pixels set in the original image, is then used to form the autocorrelation or covariance matrix (Fabbri 84). Figure 27 shows an illustration of this technique with a binary image containing several features. After the image is shifted 1, 2, or 3 pixels in each of the four 45 degree directions, the fraction of the pixels remaining is determined. From the array of values, it is easy to see that the original image has preferred orientation. Chi-squared tests against a circularly symmetric pattern can be used to quantify the probability of deviation.

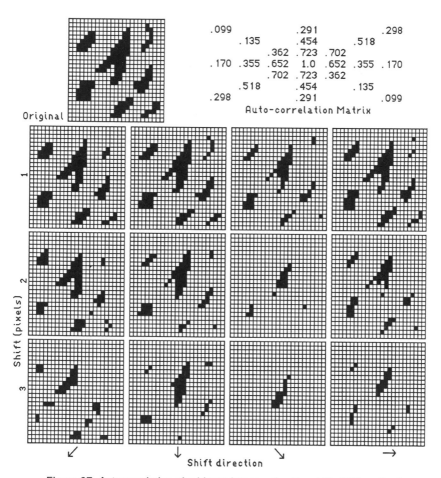

Figure 27. Autocorrelation of a binary image using the shift-*AND* method, and the resulting matrix.

Other Methods

There are other fields that measure images, often using the same computerized equipment as stereometric measurements and image processing require, but to determine the density of regions (Russ et al. 85). This is useful in autoradiography and chromatography, and some types of fluorescence microscopy. The integrated brightness of features (converted to density, dosage, or other directly useful units with an appropriate calibration curve) can be obtained from the stored image pixels. The location of the pixels may be determined by a user specified line, circle or other shape, but in some cases the same methods already described are applied to discriminate features of interest, and then the brightness values are processed automatically. The brightness, density, etc. becomes one more parameter describing the object (along with area, length, etc.) and can be used for statistical comparisons.

Particularly in the SEM measurement of particulates, it is often useful to measure not only the size and shape of features, but to make a chemical identification as well. This can often be accomplished with X-ray spectroscopy using the characteristic X-rays generated by the incident electron beam. While it is not practical to obtain a quantitative result for the particle composition (because of the irregular surface, the partial penetration of the electrons through at least the edges of the feature, and above all the very poor counting statistics due to the low number of X-rays excited during the brief moments required to obtain adequate morphological information), the pattern of elemental intensities is often quite sufficient to distinguish between the various types of materials that are of concern. Sorting the particles into bins on the basis of the intensity or intensity ratio allows the usual statistical tests to be conducted to compare particles on the basis of chemical type as well as size and shape.

It should be noted that there are a variety of non-imaging methods that provide information about particle sizes. These include X-ray diffraction (the shape of the diffraction peak broadens for very fine particles, whose effective size can be calculated from the spread); light scattering (a similar method used to determine the size of the particles in Saturn's rings); surface adhesion (the amount of an inert gas that adheres to surfaces gives a measure of the total surface area per unit weight or volume); sedimentation (particles are allowed to sink in a medium of known viscosity, and the rate of fall is used to calculate an effective size based on Stokes' law); and the Coulter counter (particles are pumped through an aperture, displacing electrolyte and changing the resistance in proportion to their volume). There are other techniques as well, many of them represented by commercially available instruments and used with success and confidence in particular industrial or research situations.

It is important to recognize that these tests may be very appropriate for the purpose to which they are put. For instance, surface adhesion tests report the amount of surface area of particles used as substrate for catalysts. The actual size and shape of the particles is largely ignored, but it is after all the surface area that controls the behavior of the catalyst. Most imaging methods of determining the surface area of the same material (for example by embedding it, cutting sections, and counting P_L) will report a different, usually lower value of surface area. This is because something different is being measured, not because one technique or the other is in error. The imaging method usually does not see fine cracks and surface irregularities that are unimportant structurally, but which are large enough to be measured at the scale of the gas molecules that adhere to the surface.

Similarly, methods that report an effective mean diameter (eg, scattering and sedimentation) are affected by the degree of surface roughness, which they cannot detect separately from size (some of the scattering methods, using coherent light, do produce a waveform signature whose Fourier transform contains shape information). But if the data obtained describe the particles in a way pertinent to their subsequent use, then the more time-consuming methods of image measurement are not justified.

A recent review of methods for particle size analysis (Barth & Sun, 85) contains exhaustive references to current work using a broad spectrum of techniques, including scattering of light and other particles and radiation, sedimentation and related chromatographic methods, and others. Microscopy and image analysis appear only briefly in this list, partially because some of the newer methods are currently producing a large number of publications as they undergo rapid development, and partly because the other methods have many applications of industrial importance. Although they give substantially less information about the nature of the individual particles, they are capable of measuring hundreds or thousands of times as many particles in the same time.

Finally, and in conclusion, it is important to understand that Stereology is a very active science now, with many mathematical discoveries still to be made which should allow new structural parameters to be defined and measured. This text has sought to present the practical side of the science: useful tools that can be put into routine use to make measurements of practical interest in many scientific fields. The people who use stereological methods are often quite distinct from those who derive them. If you encounter a new situation in which these tools seem only partially adequate, it may be possible to develop your own extensions to the theory, but your main track of research may be diverted in the meantime. The literature of stereology is extensive and widely dispersed, in different journals and even now under different subject headings. Be patient and thorough, and if possible, find a mathematician to help!

Geometric Probability

Many times in the course of this text, we have called upon the principles of geometric probability for a relationship between the probability or frequency of observation and the magnitude, for instance an area or length, of a quantity obtained by sampling particular types of objects. We will attempt here to develop the tools for making these calculations and estimations. There are two avenues by which we may approach the determinations: integration of analytic expressions, and random sampling. The former will initially be more familiar, as an outgrowth of normal analytic geometry and calculus, but the latter will ultimately be more powerful and useful for dealing with many of the problems encountered with real three dimensional features.

Methods: Analytic and Sampling

As an introductory example, consider the problem of determining the area of a circle of unit radius. Figure 1 shows the familiar analytical approach. The circle is broken into strips of width dx, whose length if expressed as a function of x is

$$L = 2 \sqrt{(1-x^2)}$$

Then the area of the circle is carried out by integration, giving

$$Area = 2 \cdot {}_{-1}\!\int^{+1} \sqrt{(1-x^2)}\ dx$$

which can be directly integrated to give

$$2 \cdot [\ 1/2 \cdot \sin^{-1}(1) - 1/2 \cdot \sin^{-1}(-1)] = \pi/2 - (-\pi/2) = \pi$$

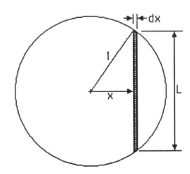

Figure 1. Integration of the area of a circle

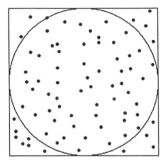

Figure 2. The "shotgun" method of measuring the circle's area.

The other approach to this problem is to draw a circle of unit radius inside a square (of side 2, and area 4). Hang the paper on a tree, back off some suitable distance, and shoot it with a shotgun. Figure 2 shows an example of what you might find. Now count the holes that are inside the circle, and those that are inside the square. The fraction that are inside the circle divided by the total number of holes should be $\pi/4$.

This is a sampling method. It carries the implicit assumption that you have a perfectly random shotgun, in the sense that the probability of holes (points) appearing within any particular unit area of the target (plane) is equal. This is in agreement with our whole idea of randomness. And, of course, the precision of our result depends on counting statistics. We will have to shoot quite a lot of holes in the target and count them to arrive at a good answer.

This is easier to do with a computer simulation than a real shotgun. Most computer systems incorporate some type of random number generator, usually a complex software routine that multiplies, adds and divides numbers to obtain difference values that mimic true random numbers. It is not our purpose here to prescribe tests of how good these "psuedo random number" generators are. (Some are quite good, producing millions or hundreds of millions of values that do not repeat and pass elaborate tests for uniformity and unpredictability; others are barely good enough for their intended purpose, usually controlling the motion of alien space invaders on the computer screen.)

Assuming that we have a random number function, and some high-level computer language in which to work, a program to shoot the holes and count them might look like this:

```
Increment = 1000
Number = 0
Count = 0
Loop:  X = RND (1)
       Y = RND (1)
       IF (X*X + Y*Y) < 1 THEN Count = Count + 1
       Number = Number + 1
       IF (Number/Increment) = INT (Number/Increment)
           THEN PRINT Number, 4*Count/Number
GOTO Loop
```

This program can be translated pretty straightforwardly into any dialect of Basic, Fortran, Pascal, etc. It uses a function *RND* that returns a real number value in the range *0 <= value < 1*, presumably with a random distribution. This means that we are in effect looking not at the entire target, but at one-quarter of it (the center is at *X=0,Y=0*). Each generated point is checked, and counted if it is inside the circle. The program prints out the area estimate periodically, in the example every thousand points, and continues until you stop it.

A typical run of the program produced the example data shown in Figure 3. Other runs, using a different set of random numbers, would be different, at least in the early stages. But given enough trials, the value approaches the correct one.

Because of the use of random numbers to measure probability, this sampling approach is called the Monte-Carlo method (after the famous gambling casino). It finds many applications in sampling of processes where the individual rules and physical principles are well known, but the overall process is complex and the steps are not easily combined in analytical expressions that can be summed or integrated, such as scattering or electrons, photons or alpha-particles.

In the example here, it is easy to see that the method does eventually approach the "right" answer, but that it takes a lot more work than the straightforward integration of the circle. However, given some other much more complex shape bounded by lines that could individually be described by relationships in *X* and *Y*, but for which the integral cannot so readily be evaluated, the Monte-Carlo approach could be a useful alternative way to find the area. We will find it especially so for three-dimensional bodies.

If the sampling pattern were not random, but instead the program had two loops that varied *X* and *Y* through a regular pattern, for instance

```
FOR Y = 0 TO 1 STEP 0.02: FOR X = 0 TO 1 STEP 0.02
```

then 2500 values would be sampled, with a good result. This "ordered" sampling is in effect a numerical integration, if the number of steps is large enough. But with ordered sampling, you have the same problems in establishing the limits as for analytic integration. Furthermore, there is no valid answer at all until you have completed the program, and no way to improve the answer by running for a longer time (if you repeat it, you will get exactly the same answer). The random sampling

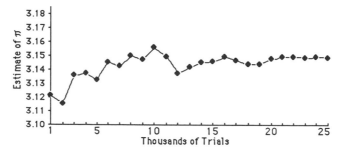

Figure 3. Typical result of running the program to measure circle area.

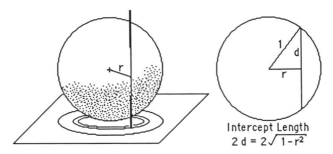

Figure 4. Diagram for intercept lengths in a sphere.

method produces a reasonable answer (whose accuracy can be estimated statistically) in a short time, and it will continue to improve with running time.

Sphere Intercepts

In Chapter 3, a graph was presented for the frequency distribution of the lengths of linear intercepts passing through a sphere. This is also a problem that can be solved analytically, as indicated in Figure 4. The method, which will only be outlined here, calculates the length of each intercept line passing vertically through a sphere on the basis of its distance from the origin (the figure shows that this is $2 (1 - r^2)^{1/2}$), and then takes the number of those lines as proportional to the area of the circular strip shown on the plane, which is $2\pi r\, dr$. The result is a frequency curve for the intercept length L whose shape is just the straight line that was shown before

$$d\eta = \pi / 2\, L\, dL$$

The Monte-Carlo sampling approach to this would be similar to that used before for the area of a circle. Because of the symmetry of the sphere, we can work just in one quadrant. X,Y values are generated by the *RND* function, and for each line that hits the sphere, the length is calculated. These lengths are summed in an array of 20 counters that become the frequency vs. length histogram. The program could be written as:

```
DIM Count(20)
FOR j = 1 TO 20: Count(j) = 0: NEXT j
INPUT "Number of trials= ";Number
FOR i = 1 TO Number
        X = RND: Y = RND
        rsquared = X*X + Y*Y
        IF rsquared > 1 then Skip
        Icept = SQR (1 - rsquared)
        Bin = INT (20*Icept) + 1
        Count(Bin) = Count(Bin) + 1
Skip: NEXT i
FOR Bin = 1 TO 20
        PRINT Bin, Count(Bin)
NEXT Bin
```

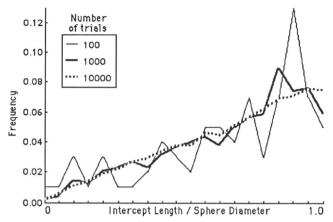

Figure 5. Probability curves for intercept lengths in a sphere generated by a Monte-Carlo program.

As before, we will not be surprised to find that a fairly large number of trials are required to obtain a good estimate of the shape of the distribution curve. Figure 5 shows three results from typical runs of the program, with 100, 1000 and 10,000 trials. The latter is a fairly good fit to the expected straight line. The uncertainty in each bin is directly predictable from the number of counts, as described in Chapter 1.

Intercept Lengths in Other Bodies

It has been possible to perform both the integration and Monte-Carlo operations very simply for a sphere, because of the high degree of symmetry. For other shapes this will be more difficult. For example, the cube shown in Figure 6 has a radically different distribution of intercept lengths than the sphere. It is very unlikely to find a short intercept for the sphere (as shown in the frequency histogram) because the line must pass very close to the edge of the sphere. But for a cube, the edges and corners provide lots of opportunities for short intercept lengths. However, if we restricted the test lines to ones vertical in the drawing, no short intercepts would be observed (nor would any long ones; all vertical intercepts would have exactly the same length). Unlike the case for the sphere, it is necessary to consider all possible

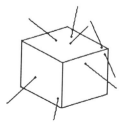

Figure 6. A cube with illustrative intercept lines.

orientations of the lines with respect to the figure. This will require more random numbers.

We have already seen that it is quite straightforward to determine a histogram of frequency versus intercept length in a circle (2-dimensional) or sphere (3-dimensional) using the Monte-Carlo approach (even though both these problems can also be solved analytically). For other shapes, the Monte-Carlo method offers many practical advantages, including that it is not difficult to program in even a small computer. However, care is required to avoid the "Bertrand's Paradox" difficulty which is discussed below, so that the intersecting lines are properly randomized in space with respect to the feature, and have uniform probability of passing through all regions and in all directions.

In two dimensions, consider the case of determining intercept lines in a square. The square may be specified as having a unit side length and corners at 0.5, 0.5; 0.5,–0.5; etc. Consider the following ways of specifying lines, and whether they will meet the requirements for random sampling (all of the random numbers are considered to vary from 0 to 1, with uniform probability density, as produced by most computer subroutines that generate psuedo-random numbers):

1) generate one number $Y = RND - 0.5$, and use it to locate a horizontal line across the square. This will uniformly sample space (if the square is subdivided into a checkerboard of smaller squares, and the number of lines passing through each is counted, the numbers will be the same within counting statistics). However, all of the intercept lengths through the square will have length exactly 1.0 because the directions are not properly randomized.

2) generate one number $THETA = \pi\,RND$, and use it to orient a line passing though the origin (center of the square). This will uniformly sample orientations, but not positions. The counts in the sampling squares near the center will be much higher than those near the periphery. This will consequently not produce a true frequency histogram.

3) generate three random numbers: $X = RND - 0.5$, $Y = RND - 0.5$, and $THETA = \pi\,RND$. The line is defined as passing through the point X,Y with slope $THETA$. It is less obvious that this also produces biased data, but it, too, favors the sampling squares near the center at the expense of those near the periphery. So do other combinations such as using four random numbers for X_1,Y_1 and X_2,Y_2 (two points to define the line).

To reveal the bias in these (and other) possible methods, it is instructive to write a small program to perform the necessary steps. This is strongly recommended to the student, as the 2-dimensional geometry and simple shape of the square make things much easier than some of the 3-dimensional problems which will be discussed shortly. Set up an array to count squares (eg. 10x10) through which the lines pass. Then use various methods to generate the lines, sum the counts, and also construct a histogram of intercept lengths. It may also be useful to keep track of the histogram of line lengths within the circumscribed unit circle around the square, since the shape of

112	110	122	143	147	163	121	114	97	103
123	170	163	181	195	175	185	153	138	124
134	191	228	242	216	220	188	189	176	157
159	181	213	259	260	255	255	247	175	130
147	172	201	253	285	287	273	254	200	166
156	175	211	259	288	305	286	258	224	151
139	183	230	252	283	265	271	270	209	172
127	165	230	252	261	230	244	193	189	160
123	158	166	190	209	207	208	179	141	125
115	126	118	127	145	159	160	132	124	118

Figure 7. Array of counts for 10x10 grid using four random numbers to generate $X_1 = RND - 0.5, Y_1 = RND - 0.5$ and $X_2 = RND - 0.5, Y_2 = RND - 0.5$ points to define the line. Note the center weighting.

that distribution is known and departures will reflect some fault in the randomness of the line generation routine.

When this is done for some of the methods described above, count arrays like those shown in Figures 7 and 8 indicate improper sampling. Both of these methods, and the others mentioned, also produce quite wrong histograms of intercept length for both the cube and sphere.

A proper method for generating randomized lines is:

Generate $R=0.5$ *RND* and *THETA* $= \pi RND$. This defines a point within the unit circle, and a vector from the origin to that point. Pass the line for intercept measurement through the point, perpendicular to the line.

(Note that generating an intercept line from two points on the circumscribed circle using two angles *THETA* $= \pi RND$ does not produce random lines, although for a sphere it does.)

An equivalent way to consider the problem is to imagine that instead of the square being fixed, and the lines oriented at all angles, the square is rotated inside the circle (requiring one random number to specify the orientation angle). Then the lines can all be parallel, just as for the earlier method of obtaining intercepts through the

177	109	59	48	39	37	37	50	63	185
197	184	173	167	142	132	138	152	161	168
235	232	216	203	213	206	239	279	297	241
249	244	282	367	312	301	333	302	266	240
257	280	318	383	423	419	341	288	260	225
220	250	298	353	422	458	406	313	240	239
196	233	280	322	360	356	361	342	274	246
197	234	225	245	222	245	399	269	242	222
201	233	181	141	128	134	145	209	196	218
181	92	51	37	35	33	38	50	101	197

Figure 8. Array of counts for 10x10 grid using two random numbers to generate points along left and right edges of square. In addition to center weighting note the sparse counts along top and bottom edges.

circle. This also requires one random number (the position along the axis). With these methods, or others which are equivalent once rotation of coordinates is carried out, proper sampling of the grids is achieved.

The result is a histogram of intercept lengths in the square as shown below. This shows a flat shelf for short lengths, where the line passes through two adjacent sides of the square. The peak corresponds to the length of the square's edge, and the probability then drops off rapidly to the maximum intercept (the diagonal, which is equal to the circle diameter).

The program used to generate the data for histogram is shown below (it uses the first of the randomizing methods described above).

```
10      DIM CT(20):DG = SQR(2)/2: P2 = 8*ATN(1)
20      INPUT "NUMBER OF LINES: ";NU
30      FOR N = 1 TO NU
40          R = DG*RND: TH = P2*RND
50          M = -1/TAN(TH)
55          B = .5 + R*SIN(TH) - M*(.5+R*COS(TH))
60          XL = 0: YL = B
65          IF YL>1 OR YL<0 THEN
                            YL=(YL>1): XL=(YL-B)/M
70          XR = 1: YR = M + B
75          IF YR>1 OR YR<0 THEN
                            YR=(YR>1): XR=(YR-B)/M
80          IF YL=YR THEN 110
90          L = INT(0.5+10*SQR((XR-XL)*
                            (XR-XL)+(X-XL)*(XR-XL))/DG)
100             CT(L) = CT(L)+1
110     NEXT N: END
```

In line 10, the array is defined (to build a 20 point histogram in this example). *DG* is the radius of the circumscribed circle, or half the square's diagonal; *P2* is 2π. In line 40, the random function is used to generate a point. Lines 50 and 55 calculate the slope (*M*) and intercept (B) of the line through this point perpendicular to the vector from the origin (the equation of the line is $y = Mx+B$). Line 60 finds the intersection of this point with the left side of the square, or if it does not pass through

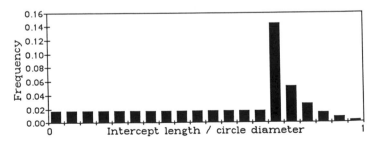

Figure 9. Histogram of intercept lengths in a square, generated by Monte-Carlo program.

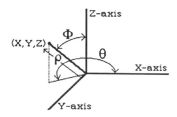

Figure 10. Angles as described in the text.

the side, the intersection with the top or bottom. The Boolean expression *YL =
(YL>1)* in line 65 is understood to produce a value of 1 if *(YL>1)* is true and 0 if it
is not. Some languages may require a more elaborate statement of this step. Line 70
does the same for the right edge, and line 80 skips lines that miss the square
altogether. Then line 90 gets the line length from the Pythagorean theorem, and
produces an integer from the ratio of this length to the circle diameter. Line 100 sums
the totals. Some additional program statements to plot or print the results would be
needed to make use of the data, but these vary from computer to computer.

Progressing to three dimensions, things become more complicated, and there
are many more ways to bias the sampling. An extension of the methods described
above for the circle and square can be used for a sphere and cube, as follows (refer to
Figure 10).

1) Generate a random vector from the origin. This requires a radius R which is
   just the circle radius times a random number, and two angles. Using the
   nomenclature of the figure above, the θ angle can be generated as $2\pi \, RND$,
   but the Φ angle cannot be just $(\pi/2) \, RND$. If this were done, the vectors that
   pointed nearly vertically would be much denser in space than those that were
   nearly horizontal, because for each vertical angle there would be the same
   number of vectors, and the circumference of the "latitude" circles decreases
   toward the pole. The proper uniform sampling of orientations requires that Φ
   be generated as the angle whose sine is a random number from *0 to 1*
   (producing angles from *0* to $\pi/2$ as desired). Through the point at the end of
   the vector defined by these two angles and radius, pass a plane perpendicular
   to the vector. Then within this plane, generate a random line by the method
   described above (angle and radius from the initial point to define another
   point, and a line through that point perpendicular to the vector from the initial
   point). This is less cumbersome than it sounds if proper use of matrix
   arithmetic is used to deal with the vectors.

2) Locate two random points on the circumscribed sphere. As in the method
   above, the points are defined by two angles, θ and Φ, which are generated as
   $\theta = 2\pi \, RND$, $\Phi = $ arc sin (RND). Connect these points with a line.

As before, alternate methods where the cube rotates while the line orientation
stays fixed in space can be used instead (but the same method for determining the tilt
angle is required). Once the line has been determined, it is straightforward to find the

intersections with the cube faces and obtain the intercept length. In the program listed below, this is done by setting up a matrix equation and solving it. Simpler methods can be used for the cube, but this more general approach is desirable when we next turn our attention to less easy shapes.

```
10      DIM A(3,4),F(6,4),LX(50)
15      P2=8*ATN(1): R=SQR(3)/2
20      FOR J=1 TO 6: FOR K=1 TO 4
25      READ F(J,K): NEXT K: NEXT J
30      DATA 1,0,0,.5,   1,0,0,.-5,   0,1,0,.5,
             0,1,0,-.5,   0,0,1,.5,   0,0,1,-.5
40      INPUT "NUMBER OF LINES: ";NU
50      FOR N = 1 TO NU
55              TH=P2*RND: E=-1+2*RND
60              E=ATN(E/SQR(1-E*E))
65              X1=R*COS(E)*COS(TH)
70              Y1=R*COS(E)*SIN(TH)
75              Z1=R*SIN(E)
80              TH=P2*RND: E=-1+2*RND
85              E=ATN(E/SQR(1-E*E))
90              X2=R*COS(E)*COS(TH)
95              Y2=R*COS(E)*SIN(TH)
100             Z2=R*SIN(E)
105             A(1,1)=Y1*Z1-Y2*Z1
110             A(1,2)=Z1*X2-Z2*X1
115             A(1,3)=X1*Y2-X2*Y1
120             A(1,4)=0
125             A(2,1)=Y1*Z2-Y2*Z1
130             A(2,2)=Z1*X2-Z2*X1+Z2-Z1
135             A(2,3)=X1*Y2-X2*Y1+Y1-Y2
140             A(2,4)=Y1*Z2-Y2*Z1
145             PC=0: FOR J=1 TO 6: FOR K=1 TO 4
150             A(3,K)=F(J,K): NEXT K
155             DE=A(1,1)*(A(2,2)*A(3,3)-A(2,3)*A(3,2))
                    +A(1,2)*(A(2,3)*A(3,1)-A(2,1)*A(3,3))
                    +A(1,3)*(A(2,1)*A(3,2)-A(2,2)*A(3,1))
160             IF DE=0 THEN 210
165             X=(A(1,4)*(A(2,2)*A(3,3)-A(2,3)*A(3,2))
                    +A(1,2)*(A(2,3)*A(3,4)-A(2,4)*A(3,3))
                    +A(1,3)*(A(2,4)*A(3,2)-A(2,2)*A(3,4)))/DE
170             Y=(A(1,1)*(A(2,4)*A(3,3)-A(2,3)*A(3,4))
                    +A(1,4)*(A(2,3)*A(3,1)-A(2,4)*A(3,3))
                    +A(1,3)*(A(2,1)*A(3,4)-A(2,4)*A(3,1)))/DE
175             Z=(A(1,1)*(A(2,2)*A(3,4)-A(2,4)*A(3,2))
                    +A(1,2)*(A(2,4)*A(3,1)-A(2,1)*A(3,4))
                    +A(1,4)*(A(2,1)*A(3,2)-A(2,2)*A(3,1)))/DE
180             IF ABS(X)>0.5 OR ABS(Y)>0.5
                        OR ABS(Z)>0.5 THEN 210
190             IF PC=0 THEN PC=1: X0=X:
                        Y0=Y: Z0=Z:GOTO 210
200             LE=SQR((X-X0)*(X-X0)+(Y-Y0)*(Y-Y0)+
                        (Z-Z0)*(Z-Z0))
205             H=INT(25*LE/R): LX(H)=LX(H)+1
210     NEXT J: END
```

In this program, line 10 dimensions an array for the histogram and line defines 2π and the radius of the circumscribed sphere. Lines 20 – 30 define the equations for each of the six faces on the cube (equations such as $1X+0Y+0Z=0.5$). Lines 55 – 75 generate a random point on the sphere. For convenience, the elevation angle E used here is the complement of the angle Φ discussed above, and the complicated ATN function is used to get the arc sine, which many languages do not provide. The point X_1,Y_1,Z_1 (and the second point X_2,Y_2,Z_2 generated in lines 80 – 100) define the intersections of the random line on the sphere.

Lines 105-120 place coefficients in the A matrix which define this line. Line 145 checks each face for intersections with the line, by placing the face coordinates into the matrix. PC is a counter which will be used to find two intersections of the line with faces. Line 155 calculates the determinant of the matrix (if zero, we skip to another face, since the current one is exactly parallel to the line). Lines 165-175 calculate the coordinates (X,Y,Z) of the intersection of the line with the face plane, and line 180 checks to see if the intersection lies within the face of the cube. For the first such intersection, line 190 saves the coordinates and increments the counter PC. When the second point is found, line 200 calculates the distance between them (the intercept length), converts it to an integer for the histogram address, and adds one count. The process repeats for the selected number of intersection lines. Again, some output or storage of the data or a graph are needed to make the data accessible.

Several additional features can be added to this type of program to serve as useful checks on whether gross errors are present in the various parts of the program. First, the sphere intercept length can also be used to build a histogram (this is just the distance between the points X_1,Y_1,Z_1 and X_2,Y_2,Z_2) which should agree with the expected frequency histogram for a sphere. Also, by summing the lengths of all the lines in both the sphere and cube, the volume fraction V_V and the surface area per unit volume S_V of the cube in the sphere can be obtained, as defined in chapter 3, and compared to the known correct values. The mean intercept length in the feature L_3 can also be obtained, which as given in chapter 3 should equal 4 V/S for the cube. It is also relatively simple to add a counter for each face of the cube, and compare the number of intersections of lines with each face (if random orientations are correctly used, these values should be the same within counting variability).

Before looking at the results from this program, It is worth pointing out that it may be relatively easily extended to other shapes, even non-regular or concave ones. The number of faces will vary, and so some of the dimensions will change, and it may be worthwhile to directly compute the equations of each face plane from vertex points. The most significant change is in the test to see if the intersection point of the line with each face plane lies within the face. For the general polyhedron, the faces are polygons. The simplest way to proceed is to divide the face into triangles formed by two adjacent vertices and the intersection point, and sum the triangle areas. If this value exceeds the known area of the face, then the point lies outside the face (provision for finite numerical accuracy within the computer must be made in the equality test).

Applying this program to a series of regular polyhedra produces the results shown in Figure 11. The results are interesting in several respects. For instance, note that whereas for the sphere, the most likely intercept length is the maximum (equal to

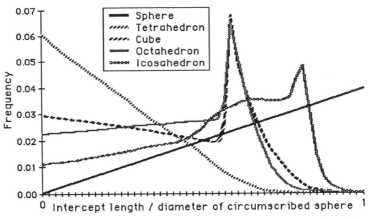

Figure 11. Frequency histograms for intercept lengths through various regular polyhedra, compared to that for a sphere, generated by Monte-Carlo program using 50 intervals for length, and about 40,000 intercept lines.

the sphere diameter), this value is quite unlikely for the polyhedra because it requires a perfect vertex-to-vertex alignment (and is impossible for the tetrahedron because even this distance is less than the sphere diameter). Also, the peaks present in the frequency histogram correspond to the distance between parallel faces (again, there are none for the tetrahedron), while the sloping but linear shelves correspond to intersections through adjacent faces.

The mean intercept lengths in the various solids are obtained from the distributions. They are listed below as fractions of the diameter of the circumscribed sphere.

Solid	Mean Intercept
Tetrahedron	0.232
Cube	0.393
Octahedron	0.399
Icosahedron	0.538
Sphere	0.667

Not all distribution curves are for intercept lengths, of course. It is also possible to model using geometric probability (or to measure on real samples) the areas of intersection profiles cut by planar sections through any shape feature. Figure 12 shows a plot (Hull & Houk 58) of the area of profiles cut through spheres and cubes. It is quite evident that the cube's corners can produce a large number of small area intersections, and that there are ways to cut areas quite a bit larger than the most probable (which is nearly equal to the area of a square face of the cube). Generating this distribution using geometric probability, or the distribution for line intercept lengths or areas in other shapes, is an excellent practical self-test of your ability to use these concepts.

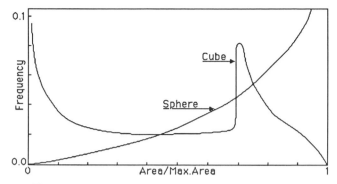

Figure 12. Intercept area distributions for a sphere and cube.

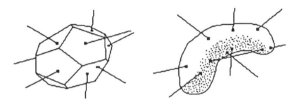

Figure 13. Illustrative intercept lines for arbitrary shapes. Concave shapes can produce multiple intercepts on one line.

For other shapes with even less symmetry, as shown in Figure 13, the analytical approach becomes completely impractical. Even Monte-Carlo approaches are time-consuming, because of the problems of calculating the intercept length, and it may take a very large number of trials to get enough counting statistics in the low-probability regions of the frequency histogram to adequately define it (particularly when the shape of the curve is to be used to deconvolute distributions from many different feature sizes).

Bertrand's Paradox

There is a problem which has been mentioned briefly in the previous examples, but which often becomes important when dealing with ways to orient complex shapes using random numbers. It is easy to bias (or, in mathematical terminology, to constrain) the orientations so that the random numbers do not produce a random sampling of the space inside the object. In a trivial example, if we only rotated the cube around one axis, we would not see all intercept lengths; but if we rotate it uniformly around all three space angles, the proper results are obtained.

A classic example of the problem of (im)proper sampling is stated in a famous paradox presented by Bertrand in 1899. The problem to be solved is stated as follows: What is the probability that a random line passing through a circle will have an intercept length greater than the side of the inscribed equilateral triangle in the circle?

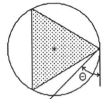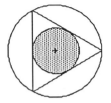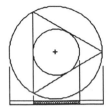

Figure 14. Illustrations for Bertrand's Paradox.

Using the three drawings in Figure 14 (and working from left to right), here are three arguments that can be presented to quickly arrive at an answer.

1. Since the circle has perfect symmetry, and any point on the periphery is the same as any other, we will consider without loss of generality, lines that enter the circle at one corner of the triangle, but at any angle. Since the total range of the angle theta is from 0 to 180 degrees, and since the triangle subtends an angle of 60 degrees (and any line lying within the shaded region clearly has an intercept length greater than the length of the side of the triangle), the probability that the intercept line length is greater than the side of the triangle is just 60/180, or one-third.

2. Since each line passing through the circle produces a line segment which must have a center point, which will lie within the circle, we can reduce the problem to determining how many of the line segments have midpoints that lie within the shaded circle inscribed in the triangle (these lines will have lengths longer than the side of the triangle, while any line whose midpoint is outside the circle will have a shorter length). Since the area of the small circle is just one-fourth the area of the large one, the probability that the intercept line length is greater than the side of the triangle is one-fourth.

3. Since the circle is symmetric, we lose no generality in considering only vertical lines. Any vertical line which passes through the circle but passes outside of the smaller circle inscribed in the triangle will have a length shorter than the side of the triangle, so the probability that the intercept line length is greater than the side of the triangle is the ratio of diameters of the circles, or one-half.

These three arguments are all, at least on the surface, plausible. But they produce three different answers. Perhaps we should choose one-third, since it is in the middle and hence "safer" than the two extreme answers!

No, the correct answer is one-half. The other two arguments are invalid because they impose constraints on the random lines. Lines spaced at equal angle steps all passing through one point on the circle's periphery do not uniformly sample it, but are closer together at small and large angles. This under-represents the number of lines that lie within the shaded triangle, and hence produces too low an answer.

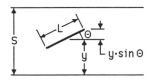

Figure 15. Geometry for the Buffon needle problem

Likewise the lines whose midpoints lie within the shaded inner circle would be correct if only one angular orientation were considered (it would then be equivalent to the third argument), but when all angles are considered it undercounts the lines that pass outside the shaded circle. It can be very tricky to find subtle forms of this kind of bias. They may arise in either the analytical integrations or the random sampling methods.

The method used to generate random orientation of lines in space for the Monte-Carlo routines described above must avoid this problem. The trick is to use $\theta = 2\pi\,RND$ and Φ = arc sin (RND), using the terminology of Figure 10. This avoids clustering many vectors around the Z axis, and gives uniform sampling.

The Buffon Needle Problem

A good example of the role of angles in geometrical probability is the problem described in Chapter 3: If a needle of length L is dropped at random onto a grid of parallel lines with spacing S, what is the probability that it crosses a line? This is another example of a problem easy enough to be solved by either analytical means or random sampling. First, we will look at the procedure of integration.

Figure 15 shows the situation. Two variables, y and *theta*, are involved; *theta* is the angle of the needle with respect to the line direction, and y is the distance from the midpoint of the needle to a line. It is only necessary to consider the angle range from 0 to $\pi/2$, because of symmetry. At any angle, the needle subtends a distance of $L \sin$ Á perpendicular to the lines, so if y is less than half this, there will be an intersection. Hence, the probability of intersection is, as was stated before

$$\int_0^{\pi/2} (L{\cdot}sin\ \theta\,/S)\ d\theta\ /\int_0^{\pi/2} d\theta = L/S[\,-cos\,(\pi/2) + cos\,(0)\,]\,/\,(\pi/2) = 2L/\pi S$$

A Monte-Carlo approach to solving this problem is shown below. L and S have been arbitrarily set to 1, so the result for *number / count* should just be $\pi/2$.

```
INPUT "How many trials: "; Number
HalfPi = 2 * Arctan (1) { to get Pi/2 }
FOR j = 1 TO Number
      Y= RND (1)
      Theta = HalfPi * RND (1)
      Vert = Sin (Theta) / 2
      IF ((Y-Vert < 0) OR (Y+Vert > 1))
            THEN Count = Count + 1
NEXT j
PRINT Number/Count
```

One particular sequence of running this program ten times, for 1000 trials each, gave a mean and standard deviation of 1.5727 ± 0.0254, which is quite respectable. There is a probably apocryphal story about an eastern European mathematician who claimed that he dropped a needle 5 cm long onto lines 6 cm apart, and counted 226 crossing events in 355 trials. He used this to estimate π, as 2 x 355 / 266 = 3.1415929 (correct through the first 6 decimal places).

Can you estimate the probability of his telling the truth about this? (Dropping the needle one more time, regardless of the outcome, would have seriously degraded the "accuracy" of the result!). This tale may serve as a caution to always consider the statistical precision of results generated by a sampling method such as Monte-Carlo simulation, as well as the statistics of measured data to which it is to be compared.

Appendix B
Representative Problems

Practically speaking, it is impossible to learn stereological principles and become accustomed to thinking in terms of three-dimensional bodies without working some examples. The ones that follow are typical of the kinds of problems that the student should do for exercise. The comments for each contain hints on appropriate methods of solution, and describe what points the problem addresses.

Problem #1. You have been measuring the size of spots on the wings of adult males of the nocturnal moth *V.Cryptica,* in North Carolina and Indiana. The data are listed below in mm (the moths are hard to catch, so you didn't get a lot of data). If you can convince your advisor that the spots show the two populations of moths to be different, you might convince him that they are separate subspecies, in which case you can finish your thesis and get a degree. How confident should you be? In other words, find ways to compare the data sets below to each other, and comment on whether they appear to represent a single population or two separate ones. If the above description does not appeal to you, pretend that the data are a) the sizes of precipitates in a Titanium alloy, after different heat treatments; b) diameters of particles used to make a ceramic, which appear to show different packing and sintering behavior; c) radii of tips on machine tool bits after use under different conditions; d) the elapsed time in seconds required to dissolve two different manufacturers' samples of a polymer in a dilute acid. Document whatever calculations you perform.

Indiana: 3.7, 3.9, 4.0, 4.2, 4.3, 4.4, 4.4, 4.6, 4.8, 4.9, 5.1, 5.3, 5.3, 5.5, 5.6

North Carolina: 4.1, 4.2, 4.2, 4.7, 5.3, 5.5, 5.5, 6.0, 6.2, 6.3, 6.7, 6.9

Comments: Getting the mean and standard deviation of each set and using an analysis of variance will indicate that these are different data sets. But neither collection of data is reliably Gaussian when checked with a chi-squared test; if you put all of the data into a single histogram and check it with a chi-squared test, the combined data set are just close to a "normal" distribution as either one is alone. For such small data sets, a nonparametric test is best. The Wilcoxon test also says that they are separate populations. If you go through all of these steps, most of the statistical calculations in Chapter 1 are involved. This should hopefully serve as a refresher for some previous statistics course. The description of the problem simply serves to remind the student that statistical tests in stereology are no different than in any other discipline.

Problem #2. Use a photocopy of Figure 1 in Chapter 1. Draw random lines and count intersections with the grain boundaries. Group your data into at least 5 sets, and calculate the mean and standard deviation of your answer. Compare this to that predicted by counting statistics, and to the "true" answer given in the text.

Comments: This is a good first introduction to drawing lines, counting, and performing simple statistical calculations. The global average of results from the entire class should be better than any single value.

Problem #3. First, make an eyeball guess at the area fraction of each image that is covered by the black phase. Rank them in order. Then perform a measurement of area fraction using an area, line or point count method. Compare the results.

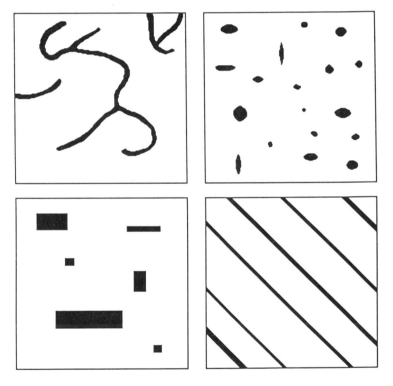

Problem #3

Comments: The human eye is notoriously poor as an estimator of area fraction, and it is good for students to accept this quickly. Point counts using the reticle images in Chapter 3 (which can be photocopied onto transparent overhead material for the purpose) are the most efficient way to get area fractions, but these images have fairly close results so that a large enough number of points to get good statistical precision must be used. The strongly anisotropic nature of two images will strongly bias the results if non-random placement of points or lines is used. Cutting and weighing is actually a pretty good way to do this problem, because there are not too many features, and they are not too complex.

Problem #4. Draw at least 10 random sections (ie. plane sections at random orientations) through a chain of the type shown in the figure.

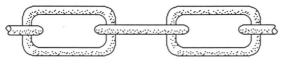

A typical link chain in a very non-random orientation

Problem #4

Comment: This kind of problem encourages students to think three-dimensionally. It is very easy to fall into the trap of lining the sections up with one or another of the principal directions of the chain and thus overlook some important section appearances. It is also instructive to collect a few dozen random intersections together and ask students to attempt to visualize what kind of three-dimensional object could have produced them. (They look much like the tracks of some strange bird in the snow.)

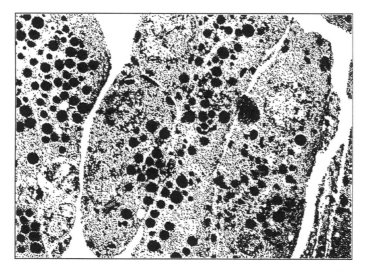

Problem #5

Problem #5. For the idealized transmission electron microscope image shown in Chapter 2, Figure 7, determine the volume fraction of the stained organelles in the cells and in the cytoplasm (the portion of the cell outside the nucleus), and the mean diameter of the organelles, assuming they are spherical. Then carry out the same operation on the real image shown (a transmission electron micrograph of rat pancreas, prepared by cryo-sectioning and post-stained with uranyl acetate). For each image, use 30 μm as the image width.

Comment: Identifying and selecting the lysosomes is much more difficult in the real image, as are the inter-cellular spaces and the nuclei. This is only partly due to the printed quality of the image. In the section there are several other dark structures, and the preparation poorly outlines the nuclei. Of course, the fraction of the image occupied by lysosomes, nuclei, and cells must all be determined. This section is actually very thin (<80 nm); what would change in the calculations if it were 1 μm thick?

Problem #6. Given a distribution of spherical particles as shown in the figure, write and run a Monte-Carlo program to generate a histogram of intercept lengths through the particles, with at least a few thousand total intercepts (use at least 30 steps for intercept lengths from 0 to 3 μm). Then deconvolute the resulting histogram using the method shown in Chapter 4.

Comment: Besides getting useful practice in performing this type of simulation (especially if Appendix A is not used in the course), the student will have a chance to deal with a real histogram containing much more data than could be practically measured by hand, and will encounter the problem of finding negative values for the frequency of some diameters. The magnitude of the values for diameters less than 2, which were not present in the original distribution, will demonstrate the propagation of errors.

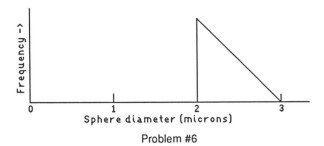

Problem #6

Problem #7. Write a Monte-Carlo program to get the intercept length distribution, or the intercept area distribution, of a right circular cylinder with length equal to twice its diameter. Run the program and show the histogram; report the mean intercept length, and the mean caliper diameter.

Comment: Many other shapes can be substituted, such as ellipsoids of revolution, cylinders with other aspect ratios, or polyhedra. Getting the method right is as important as obtaining the right result. Any computer to which the student has access, and any language, are acceptable. Learning to randomize lines in space, and visualizing the orientations and intersections of lines or planes with objects, are very important steps in coming to an understanding of stereological principles. Getting the program to run, and to generate the correct results, takes a lot of time, so this can be spread over several weeks.

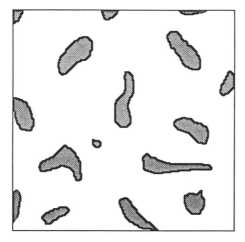

Problem #8

Problem #8. For the image shown (a section through a material containing a dispersed phase, or through a cell containing separated organelles), find the mean free distance between the particles. Then treat the matrix as a membrane separating the black particles and determine its mean thickness.

Comment: Measuring the mean free spacing as described in Chapter 3, and then using the method shown in Chapter 4 for handling lamellae, seem at first like rather different operations. The two results should, of course, be the same.

Problem #9. On a polished section through a specimen, dispersed features appear as separated profiles of long, narrow cross section. From independent sources of information, you know that these are really thin disks, or platelets (as opposed to rods). You do NOT have the ability to prepare additional polished sections. The measurements which are available to you are: a) linear intercept counts (number of points per line) for lines drawn randomly or in an specific direction; and b) the area, perimeter, length (longest Feret's diameter), breadth (shortest Feret's diameter) and angle of each individual feature. Describe a procedure, with all relevant equations, tables and coefficients to characterize the following material properties: 1) the degree of preferred orientation in the microstructure and the statistical probability that it is significant; 2) the mean free distance between particles; 3) the mean size (caliper diameter); and 4) the number of particles per unit volume.

After completing the outline, apply the method(s) you have chosen to the image shown. Note that there are not really enough features to do a statistically satisfactory job, so this is a demonstration rather than an experiment. But can you estimate how many such photos would be required to obtain results with about 10% expected error?

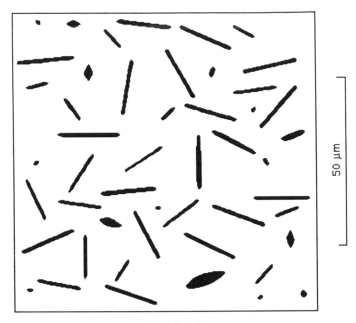

50 μm

Problem #9

Comment: There are several ways to obtain the required information. For instance, preferred orientation can be determined from a plot of number of particles versus orientation angle, or from a plot of number of intercepts per line length in various directions, or from a calculation using intercept counts in just two directions. The latter requires the least work, but the rose or orientation plots are more readily interpreted from a statistical point of view. Also the number of additional photos required to achieve a 10% precision is different for different measurements; for instance, many more images would be needed to establish the mean diameter to within 10% than the number of particles per unit volume. Using the relationship in Chapter 4 $(N_A=N_V d)$ is a more efficient way to get d than constructing a cumulative histogram to find the mean. This problem exercises a number of the common measurement methods and requires some statistical thinking to evaluate the result.

Problem #10. Draw representative projected images of cubic particles in random orientations with a) a uniform size of 3 μm; b) a uniform size distribution from 1 to 5 μm From the area, length and perimeter of these shadow images, construct equations which estimate the volume of the particles. Apply your method to arrive at an estimate for the mean particle size.

Comment: Alternately, a computer program can be written to generate the images. Even if the fact that the particles are cubes is known, it is not simple to estimate the volumes accurately from the dimensions of the shadows. Models other than the ones presented in Chapter 6 may be evaluated. The "mean size" is not specified in the problem statement, but it is implied that it should be the mean based on volume, which would not be the same for the two different cases.

Problem #11. Spherical particles, all opaque and exactly 1 μm in diameter, are embedded sparsely and randomly in a transparent matrix. Sections cut with finite thicknesses of 0.1, 1 and 5 μm are viewed in transmission. Draw schematically the appearance of a histogram of frequency versus diameter of the circles that are seen in the projected image. Clearly mark any differences that are present in the three histograms, and explain briefly why they appear. Optionally, you may write a computer program to generate simulated histograms for the three cases.

Comment: The histogram shape for an infinitely thin slice is well known, and for an infinitely thick slice is very simple (all circles have the same size equal to the sphere diameter). It is not hard to predict the effect of finite thickness, and this gives good experience in dealing with real microscope samples which usually have some thickness (even for polished surfaces, which have some relief).

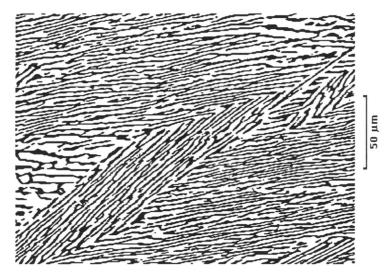

Problem #12

Problem #12. For the image shown (pearlite, a eutectic consisting of iron and iron carbide in steels), construct a rose of intercept lengths. Calculate the degree of preferred orientation.

Comment: The method is straightforward, and this gives good practice in handling intercept counts, which are one of the most efficient ways to perform manual measurements.

Problem #13. Construct a Richardson plot for the image of the leaf shown in the figure. Over what range of dimensions can it be described as having a fractal shape?

Problem #13

Comment: Students should be challenged to find other natural shapes for this type of measurement, as well. Discuss the parameters of real objects which have a fractal aspect, or aspects which correlate with fractal dimensions.

Problem #14. Two sets of stereo pair photomicrographs of a hardness indentation in aluminum were made with an SEM. For each picture, the total image height is about 120 μm. The angle of the nominal surface is 0 and 10 degrees in the top pair, and 45 and 55 degrees in the bottom pair. From measurements on each pair, independently determine the depth of the indentation below the surface. Note: the surface immediately around the edge of the indentation is raised somewhat by displaced material; pick points to represent the surface which are away from the edge. From the depth values, calculate the volume of the indentation.

Comment: The two measurements should, of course, give the same result (about 20 μm) for the depth. Since the absolute angles are given, it is possible to simplify the measurement by using two surface points along a line parallel to the tilt axis to measure the parallax of the point at the peak of the pyramid, but a more general approach will also determine the surface angle. Common errors include measuring the depth of the central point in a vertical direction rather than perpendicular to the plane, and measuring the length of the side of the indentation in a horizontal direction instead on parallel to the plane. The pictures are taken from the paper by Roberts and Page (81).

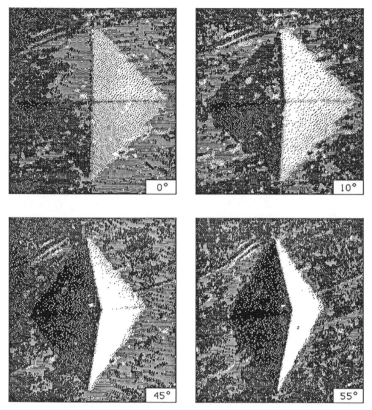

Problem #14

Problem #15. Stereo pair photographs show two points which are actually 10 μm apart and with a true elevation difference of 10 μm relative to the nominal specimen surface. They are oriented at an angle of 45 degrees to the tilt axis used for the stereo pair photos. Determine the expected error in the height measurement in each of the following circumstances:

a) the left photo is at a nominal tilt of zero and the right photo is at 5 degrees. Both photos are at a magnification of 100X. The uncertainty in determining the location of the points (eg. with the stylus of a digitizing pad or a vernier scale) is 0.2 mm.

b) as in a) except that the right photo is at a tilt of 15 degrees instead of 5 degrees.

c) as in a) except that the photos are at tilts of 40 and 50 degrees.

For d) through g) the left photo is at a tilt of zero and the right photo is at a tilt of 10 degrees. Stylus positioning error is negligible.

>   d) the alignment of the tilt axis is 5 degrees off from the direction that is
>   assumed in measuring the photos.
>   e) the tilt is actually 9 degrees instead of the assumed 10 degrees.
>   f) the magnification is 90X rather than the assumed value of 100X
>   g) the magnification of the right hand picture (only) changed to 90X because
>   the working distance changed when the specimen was tilted, but this is
>   not known.

Comment: This problem can be solved either by just calculating through the equations for all cases, or applying error analysis methods to differentiate the equation for elevation. In either case, it demonstrates the magnitude of errors that are likely to be encountered in routine stereoscopy. The magnitude of the error in part (g) is by far the largest, illustrating the importance of keeping the focal length constant when taking stereo pair photos in the scanning electron microscope.

Problem #16. Sketch 4 serial sections through a complex structure such as the spiral noodle pictured below, perpendicular to a direction inclined 30 degrees with respect to its major axis, and separated by 0.5 mm.

Comment: This exercise will help the student visualize the relationships between three dimensional objects and sections through them, which can be very hard to learn. Noodles come in a wide variety of shapes and are complex enough to be excellent for this purpose. An instructive exercise is to cook them, embed them in Jello, and then section them. Since the number of features per unit volume, their dimensions, and the volume fraction are known (the volume of noodles is easy to determine by displacement of water in a graduated cylinder), photos of the cut sections can be used for a variety of measurement exercises.

Problem #16

Bibliography

This is not intended to be a complete bibliography on stereology, statistics and related subjects. In particular, few of the numerous publications on geometric probability, or those with derivations of parameters for particular types of microstructures, are cited, and only those applications from which examples have been adapted for use in this text. The books by Underwood and Weibel contain exhaustive bibliographic references in these areas, and should be consulted as required. Both have been extremely helpful in guiding the presentation here. The other significant books in the field are those of Saltykov, Elias, and Dehoff, which are included in the list below.

References to the original presentation of seminal ideas have been included (except in a few cases such as Mandelbrot where the same author has subsequently published a more complete version), and it is worth noting that quite a few have appeared in the Transactions of the AIME, and more recently in the Journal of Microscopy, which has become a focal point for much of this work. Inclusion of papers from diverse fields including biology, geology and astronomy indicate the breadth of application of these methods.

Any good textbook on statistical analysis will be a valuable supplement to Chapter 1. This text owes a particular debt to the presentation format of P. R. Bevington, *Data Reduction and Error Analysis for the Physical Sciences*, McGraw Hill, New York, 1969. The book *Geometrical Probability* by M. G. Kendall and P. A. P. Moran (Charles Griffin, London 1963) is also recommended as a comprehensive introduction to that subject.

The books by Pratt and by Rosenfeld and Kak give complete introductions to the processing of grey scale images, and deal slightly with the task of segmenting the image into regions (which are the features or objects we wish to measure). They also contain exhaustive references to the original literature, which is not cited here.

Abercrombie, M. "Estimation of nuclear population from microtomic sections," Anat. Rec. 94 (46) 239

Barth, H. G., S.-T. Sun "Particle Size Analysis," Anal. Chem. 57 (85) 151R

Beddow, J. K., G. C. Philip, A. F. Vetter "On Relating some Particle Profile Characteristics to the Profile Fourier Coefficients" Powder Technology 18 (77) 19

Boyde, A. "Quantitative photogrammetric analysis and quantitative stereoscopic analysis of SEM images," J. Microsc. 98 (73) 452

Buffon, G. L. L. "Essai d'arithmetique morale," Suppl. a l'Histoire Naturelle, Paris, 4 (1777)

Cahn, J. W., R. L. Fullman "On the use of Lineal Analysis for Obtaining Particle Size Distribution Functions in Opaque Samples," Trans. AIME 206 (56) 610

Cahn, J. W., J. W. Nutting "Transmission Quantitative Microscopy," Trans. AIME 215 (59) 526

Cahn, J. W. "The significance of average mean curvature and its determination by quantitative metallography," Trans. AIME 239 (67) 610

Cauchy, A. in "Ouvres completes d'Augustin Cauchy Vol II" Gauthier-Villans, Paris (08) 167

Chandrasekhar, S. "Stochastic Problems in Physics and Astronomy," Rev. Mod. Phys. 15 (43) 86

Coster, M., J.-L. Chermant *Precis D'Analyse D'Images*, Centre National de la Recherche Scientifique, Paris (85)

Cruz-Orive, L.-M. "Particle size-shape distributions: the general spheroid problem," J. Microsc. 107 (76) 235; 112 (78) 153

Cruz-Orive, L.-M. "Distribution-free estimation of sphere size distributions from slabs showing overprojections and truncation, with a review of previous methods," J. Microsc. 131 (83) 265

Dehoff, R. T. "The Determination of the Size Distribution of Ellipsoidal Particles from Measurements made on Random Plane Sections, " Trans. AIME 224 (62) 474

Dehoff, R. T. "The determination of the geometric properties of aggregates of constant size particles from counting measurements made on random plane sections," Trans AIME 230 (64) 617

Dehoff, R. T. "The quantitative estimate of mean surface curvature," Trans AIME 239 (67) 617

Dehoff, R. T., F. N. Rhines *Quantitative Microscopy*, McGraw Hill, New York (68)

Dehoff, R. T. "Stereological uses of the area tangent count," in Geometrical Probability and Biological Structures: Buffon's 200th Anniversary (R. E. Miles, J. Serra, ed.) Springer Verlag, Berlin (78) 99

Delesse, A. "Pour determiner la composition des roches," Ann. des Mines 13 (1848) fourth series, 379

Dorfler, G. "A system for stereometric analysis using the electron microscope," in *Stereology* (H. Elias, ed.) Springer Verlag, New York (68) 277

Ebbeson, S. O. E., D. Tang, "A method for estimating the number of cells in histological sections," J. Microsc. 84 (65) 449

Elias, H. *A Guide to Practical Stereology*, Karger, Basel (83)

Fabbri, A. G. *Image Processing of Geological Data* Van Nostrand Reinhold, New York (84)

Feret, L. R. *La grosseur des grains*, Assoc. Intern. Essais Math 2D, Zurich (31)

Fullman, R. L. "Measurement of Particle Sizes in Opaque Bodies," Trans AIME 197 (53) 447;1267

Gil, J., E. R. Weibel "Morpological study of pressure-volume hysteresis in rat lungs fixed by vascular perfusion" Respir. Physiol. 15 (72) 190

Gundersen, H. J., T. B. Jensen, R. Osterby "Distribution of membrane thickness determined by lineal analysis," J. Microsc. 113 (78) 27

Gurland, J. "The Measurement of Grain Contiquity in Two-Phase Alloys," Trans AIME 212 (58) 452

Haralick, R. M., K. Shanmugam, I. Dinstein "Textural features for image classification," IEEE Trans. Syst. Man. Cybern. vol. SMC-3 (73) 610

Heinrich, K. F. J. "Use of color scales in microanalytical maps," J. de Physique 45 (84) 201

Heyn, E. "Short reports from the Metallurgical and Metallographic Laboratory of the Royal Mechanical and Technical Testing Institute of Charlottenburg," Metallographist 6 (03) 37

Hilliard, J.E. "The calculation of the mean caliper diameter of a body for use in the analysis of the number of particles per unit volume," in *Stereology* (H. Elias, ed.) Springer Verlag, New York (68) 211

Hull, F.C., W. J. Houk "Statistical Grain Structure Studies, Plane Distribution Curves of Regular Polyhedrons," Trans AIME 197 (53) 565

Hurlbut, G. S. "An Electric Counter for Thin Section Analysis," Amer. J. Sci. 237 (39), 253

Jeffries, Z., A. H. Kline, E. B. Zimmer "Determination of Grain Size in Metals," Trans. AIME 57 (16) 596

Kaye, B. H. "Multifractal Description of a Rugged Fineparticle Profile" Particle Characterization 1 (84) 14

Koch, von, H. "Sur une courbe continue sans tangente, obtenue par une construction geometrique elementaire," Arkiv fur Matematik, Astronomie och Fysik 1 (04) 681

Laibowitz, R. B., B. B. Mandelbrot, D. E. Passoja (ed.) *Fractal Aspects of Materials* (extended abstracts), Materials Research Society, Pittsburgh (85)

Lord, G. W., T. F. Willis "Calculations of Air Bubble Size Distribution from Results of a Rosiwal Traverse of Aerated Concrete," ASTM Bulletin 177 (51) 56

Mandelbrot, B. B. *The Fractal Geometry of Nature*, Freeman, New York (83)

McMahon, T. A., J. T. Bonner *On Size and Life*, Freeman, New York (83)

Mecholsky, J. J., D. E. Passoja "Fractals and Brittle Fracture" in Laibowitz, et. al., (85) op.cit.

Merz, W. A. "Die Streckenmessung an gerichteten Strukturen im Mikroscop und ihre Anwendung zur Bestimmung von Oberflachen Volumen Relationen im Knochengewebe," Mikrosckopie 22 (67) 132

Minkowski, H. "Volumen und Oberflache" Math. Ann. 57 (03) 447

Moore, G. A. "Is Quantitative Metallography Quantitative?" in *Applications of Modern Metallographic Techniques*, STP 480, ASTM, Philadelphia (70)

Pratt, W. K. *Digital Image Processing*, Wiley, New York (78)

Richardson, L. F, "The problem of contiguity: an appendix of statistics of deadly quarrels," General Systems Yearbook 6 (61) 139

Roberts, S. G., T. F. Page "A microcomputer-based system for stereogrammetric analysis," J. Microsc. 124 (81) 77

Rosenfeld, A., A. C. Kak *Digital Picture Processing* (2nd ed.), Academic Press, New York (82)

Rosiwal, A. "Uber geometrische Gesteinsanalyysen," Verhandl. der K.-K. geologische Reichanstalt 5/6 (1898) 143

Russ, J. C., W. D. Stewart "Quantitative image measurement using a microcomputer system," Amer. Lab. 15,12 (83) 70

Russ, J. C., J. C. Russ III "Image Processing in a General Purpose Microcomputer," J. Microsc. 135 (84) 89

Russ, J. C., W. D. Stewart, J. C. Russ III "Densitometric Image Measurement," American Laboratory 17,4 (85) 41

Saltykov, S. A. *Stereometric Metallography*, Metallurgizdat, Moscow (58)

Saltykov, S. A. "The Determination of the Size Distribution of Particles in an Opaque Material from the Measurement of the Size Distribution of their Sections," Stereology, Springer-Verlag, New York (67) 163

Saxton, W. O., T. L. Koch "Interactive Image Processing: Organization," J. Microsc. 127 (82) 69

Schwarz, H, H. E. Exner "The implementation of the concept of Fractal Dimension on a Semi-Automatic Image Analyzer" Powder Technology 27 (80) 207

Serra, J. *Image Analysis and Mathematical Morphology*, Academic Press, New York (82)

Thaulow, N., E. W. White "General method for dispersing and disaggregating particulate samples for quantitative SEM and optical microscopic studies," Powder Technol. 5 (71) 377

Thomson, E. "Quantitative Microscopic Analysis," J. Geol. 38, 3 (30) 193

Tomkeieff, S. I. "Linear Intercepts, Areas and Volumes," Nature, 155 (48) 24;107

Underwood, E. E. *Quantitative Stereology*, Addison Wesley, Reading MA (70)

Weibel, E. R., D. M. Gomez "A principle for counting tissue structures on random sections," J. Appl. Physiol. 17 (62) 343

Weibel, E. R., G. S. Kistler, W. F. Scherle "Practical methods for morphometric cytology," J. Cell Biol. 30 (66) 23

Weibel, E. R. *Stereological Methods Vol. I & II*, Academic Press, London (79)

Wicksell, S. D. "The corpuscle problem," Biometrica 17 (25) 84; 18 (26) 152

Index